NORTH WRITERS

The University of Minnesota Press
gratefully acknowledges assistance provided
for the publication of this volume
by the John K. Fesler Memorial Fund.

NORTH WRITERS
A
STRONG WOODS
COLLECTION

John Henricksson, Editor

University of Minnesota Press
Minneapolis   Oxford

Published by the University of Minnesota Press
2037 University Avenue Southeast, Minneapolis, MN 55414
Printed in the United States of America on acid-free paper

Library of Congress Cataloging-in-Publication Data

North writers : a strong woods collection / edited by John Henricksson.
p.   cm.
Includes index.
ISBN 0-8166-1950-6
1. Natural history—Authorship.   2. Natural history literature.
I. Henricksson, John.
QH14.N67   1991
508—dc20   91-12035
                                                    CIP

A CIP catalog record for this book is available
from the British Library

For Julie, who shares the magic

CONTENTS

Part Two: Interaction

ACKNOWLEDGMENTS

The task of editing an anthology appears, at first glance, to be a rather sedentary and pleasurable journey through the literature of a particular genre or region. There will be time for musing over some profound thoughts, soaring passages, and exciting adventures; time for rereading old favorites and discovering new talents. There is much of that. But soon the road gets rocky and strewn with the realities of painful choices, endless searches, self-doubts, and the formation of a cast, or structure, for the collection to assume.

Fortunately, I met four very capable and insightful people along the way, each of whom gave generously of her knowledge and strengths: Carolyn Gilman, former regional editor of the University of Minnesota Press, who provided the important initial encouragement and impetus for the project; Loree Miltich, editor of *New North Times*, who shared her knowledge of Minnesota writers and her vision of north country; Meg Aerol, University of

Minnesota Press editor, who brought creativity and some badly needed discipline and organization to the process; and Patricia Hampl, whose intellectual honesty and constructive criticism were invaluable.

They all have my gratitude and admiration.

INTRODUCTION

Nature awakens an instinctive, inspired, unnameable sense—not an extrasensory sixth sense, but rather a strange combination of all sensations. For some, the earth provokes an immediate physical response; for others, the result is closer to a spiritual experience. Regardless of the reaction, this lure of the land is the genesis of nature writing.

Although our animal origins are often forgotten in the midst of modern society, nature serves as a reminder of the wildness within us. As a formal, literary genre, nature writing is unified by one element only—the earth. In fiction, poetry, or nonfiction the environment constantly remains the main character. Whether the author is Henry David Thoreau at Walden Pond or John McPhee in the high plains, nature writing describes and examines the environment.

Among the many regions of the world studied and described by nature writers in the last two hundred years, middle

America has often been disregarded. Sigurd Olson has immortalized the canoe country, and more recently, Paul Gruchow and others have revived our interest in the prairies. However, the "north woods," or boreal forest, remains relatively unknown to most readers. This forgotten region, from the northeastern tip of Minnesota to the southern shore of Lake Superior, was called the *bois forts*, or the Strong Woods, by early French explorers.

Geologists, wildlife biologists, and silviculturalists have produced volumes of scientific data from this wild laboratory. Canoeists and sports enthusiasts have written about this terrain, but it is not familiar ground to a national audience of the nature-writing genre. Helen Hoover, Sig Olson, and a few others are well known for widely read books, and the essays of Jim Dale Vickery and Peter Leschak appear frequently in national magazines. But most of the selections in this anthology are from a variety of small-press publications and local media sources; they represent the voices of men and women who live "up north" and remain continually involved with this unique homeland. Writing of solitary nights under the aurora borealis, ice-locked winters, long, misty reaches of island-studded lakes, and occasional confrontations with wolves, these authors speak to the "up north" in everyone.

Earlier residents, the Anishinabee—or Ojibway—were consummate storytellers, and perhaps the spiritual forebears of today's north writers. Their legends of creation and history should be chronicled as the first nature writing of the region. The Anishinabee on the shores of the great lake Kitchi Gami named a little spirit cedar Mando Gee Zhi more than four hundred years ago. Best known today as the Witch Tree, it still stands in a rocky cove near the Canadian border. The Anishinabee created pictographs of nature and the spirit world on rocks and cliff faces along the boundary waters between Minnesota and Ontario from Lac la Croix to the Pigeon River; generally joyful and wondrous, these pictorial legends are still visible, their sightings a highlight in the adventures of many canoeists. The Anishinabee named Mando Gee Zhi, Saganaga, and the Kawishiwi; French voyageurs gave placenames like Maligne, La Croix, and Lac Supérieur. This mixture of languages and cultures provides the backdrop for a kind of writing that speaks today to both environmental and cultural literature.

The works contained in this volume represent a progression in nature writing from solitary exploration of the land to historical documentation of cultural interaction. Without implying that either stance is more useful or appropriate, the division of the book reflects two ways both writers and readers can approach nature writing—as a personal learning experience, or as an analysis of the way entire societies fit into an ecological plan.

The pieces included in Part One are by writers who attempt to communicate the personal sensation of nature, something that modern society increasingly renders alien. The style of writing varies greatly, and selections have been chosen to offer a wide range of voices and sensibilities rather than one formulaic approach. Many of the pieces are memoiristic, at times becoming the stream-of-consciousness record of an event. Others use a more impersonal, slightly scientific style to record only the most essential information. Still others rely on traditional narrative techniques, closer to the oral tradition of the region. The journalistic entries fall between the brevity of journal entries and the thoroughness of a complete narrative. Together, these contributions present a complex picture both of the north woods and of the variety contained within the genre of nature writing itself.

The second part of the anthology explores the cultural complexity of the region. Heart Warrior Chosa, Sigurd Olson, and Michael Furtman describe the tenuous relationship between the land's original inhabitants and its much more recent settlers. In Minnesota, as in many other parts of the United States, the culture of the native people is largely misunderstood. Nature writing represents an attempt to rectify this misunderstanding, drawing links between the contemporary world and its ignorance of peoples who hold nature in the greatest esteem. The north writers trace the evolution of relations between these vastly different cultures. Both "Glimpses of the Past" by Grace Lee Nute and "Adventures in Solitude" by Calvin Rutstrum provide historical information even as they reveal the difficulty of merging conflicting perspectives in a land where it is often difficult simply to stay alive.

Ultimately, as Douglas Wood writes in his final essay, the world—once thought to be so large—is growing smaller. In a historical context, nature writing is no longer the combination of travelogue and adventure it once was. Modern technology allows us to preview, in movies and on TV, corners of the world (and even edges of space) that are not accessible to humans. The value of nature writing today lies in its ability to *translate* nature, not merely uncover it. As Wood concludes, "Sometimes beneath the stars and the shadows of trees, as the woodsmoke rises and firelight dances on a circle of faces, the feelings rise strong once more, and the world is again small, and beautiful."

NORTH
SHORE SYMPHONY

Michael Dennis Browne

Journal 1 (Charles Lanman/J. Elliot Cabot)

The waters of this magnificent lake are so marvelously clear . . .
It is like being transported to the early ages of the earth,
when the mosses and pines had just begun to cover the primeval
rock, and the animals ventured timidly forth into the new world.

North

In sleep I heard
a bird call out a name
I thought I knew

I shed
my skin
of speech

and went
wordless

into a silence
into a beginning

I went
North   North

Here where my heart
has steered a hundred times
I go
North   North

Pines, you deepen now
I push through tunnels
into a longer light

into a silence
into a beginning

to drink, to kindle
a quietness

and listening   listening

to be shore, and shaped
to be tree, trembling
to be rock, water and flower
to be flower, to flame and fade

I go
North   North

WHY I LIVE
IN THESE HILLS

Joanne Hart

I stay here for the northern sky,
for stars caught in the hairy pines,
for moons that come at will and go
without a by-your-leave. No matter
if the evening rain clouds bank
the western gap up-river, or if
thunder groans over the mountains,
northward there will be the light.
I stay here for the sky outside
these windows, for the random shapes
of trees against a brightening
and paling glow, for smell and crackle
of a northern fire I can't set
or smother. Groggy as I make
my bed and check my clock, clumsy

with boot laces, buttons, yet
I come alive my lamp shut off,
and when at last I fall to sleep,
a northern pulse beats into mine.

PART ONE
EXPLORATION

THE SEASON OF STILLNESS

Denny Olson

Winter is the real essence of the North. It defines the boundaries of life, and shapes the skills of every creature. It is long and intense. Millions of birds escape from it every autumn, preferring instead to run the terrible gauntlet of migration. Small animals find a tiny room, a refuge, and stay immobile. Bacteria shut down. Most reptiles and amphibians don't even bother to live in the North. I don't blame them. No heat.

Except for an occasional life-death dance between a wolf and a deer, or a gray jay on reconnaissance maneuvers, the lack of motion and sound is awesome—even by wilderness standards. The only sensory inputs are gray sky, dark green, and white—lots of white.

Tracks appear overnight and then get snowed under, historical records in disappearing ink. To those who leave the history, deepening snows present problems.

But, in nature's magic touch, problems and solutions

dance through time on equal footing. Hare, lynx, and wolf have outsized feet to buoy themselves. Moose travel on stilts, stepping gracefully over three-foot drifts. Shrews and voles spelunk in sub-nivian caves created by geothermal heat. Deer congregate, making escape trails by force of sheer numbers. Grouse dive into the insulating powder and explode toward the aspen tops only at feeding time.

But all of this activity is the exception. The rule is a seemingly empty landscape, with no "action." The key to seeing is slowing observations to the pace of the season. That's what the wilderness has always been about—its own terms, not ours.

Dead-looking trees, immobile and stoic, get ready for winter completely within themselves. Winter dictates the terms, and trees seal the contract in millennia of adaptations.

Spruces, with flexible branches and perfect posture, unload snow automatically. But some, by accident of birth or environment, are slightly tilted. An unusually heavy, wet snow may load the upper sides of these crooked trees, violently snapping their trunks. Their seeds, with tilted traits, will never be spread again.

The openings created by these culled "misfits" initiate a series of curious events. Surrounding spruces are tempted by sunlight, and over a few years grow toward the openings—becoming misfits themselves and meeting the same fate. This recurs until thousands of room-sized amphitheaters speckle the northern forest, each a grim testament against disobedience to the holy vertical.

Birch trees, resilient through eons of adaptation, can bend to the ground under snowload and wait for spring. Aspens weather the assault of snow with thin expendable branches, which snap off and fall to the ground, causing a flurry of night-feeding activity by snowshoe hare connoisseurs. By day the hares hide in "quamanik," an eskimo name for the hollow snow-shadow under spruces and firs.

And then there is the cold.

On January nights at thirty below, stream openings close quickly. Bogs lie lifeless—their undulating mounds of snow only hint at the shapes below. Chipmunks are curled in secret places, torpid, but safe against the angry cold sweeping down from the tundra. Winter cold grips the North like hardened mortar.

Leafless trees look like withered fingers, frozen in death, fleshless and picked clean by the wind. But they harbor secrets. Their buds are programmed to erupt into leaves, twigs, and flowers. Water is conserved by cork leaf scars and water-tight bud scales.

Yet too much conservation can be fatal. Maples store liquid in their roots, and in a January thaw, let their sap flow freely. But here, where forty below is inevitable, a steep drop in temperature freezes the flowing sap. With deafening booms, maples explode from top to bottom, dying in their youth. On a map, the "maple line: coincides with the 'forty-below line'." Most boreal trees are more successful than maples. They store water between their living cells. The water is rich sugar and starch antifreeze, and it doesn't expand even if it hardens.

And there's more. When I puncture a balsam fir bark blister at forty below, the resin flows freely. Microscopic closeable pores are recessed far into the needles because their fragility demands protection from the elements. These slim leaves have very little surface area to freeze. Each boasts a waterproof waxy cover which prevents drying. They need it. Winter humidity levels in the North rival those of a desert.

For small annual plants, life is even more tenuous. All their hopes are wrapped in seeds. It is not surprising that most plants of the North are perennials, reproducing through their roots and stems, placing little faith in courtship.

So the stories are subtle, and still. But the magic woven by plant and animals on the edge of death is not only interesting, but profound.

Behind your doors and walls, with the fireplace reminiscent of summer campfires, it's easy to be indifferent to an inanimate season outside. It's easy to forget the processes, both biological and spiritual, which created winter and yourself together. But should you ever venture there, strap on snowshoes and feel the frostbite, you become a warp in a weave, a strand in an infinite kinetic fabric, inseparable.

In any outdoor experience, this reminder of our roots is

more than a reward. We belong. And the self-conceived realities of civilization are displaced by those much bigger than us.

By seeing these larger realities we come home to nature. By sharing them, we become naturalists.

THE WILDNESS WITHIN

Douglas Wood

Perhaps our best hope for preserving the wild world, and our Earth as a whole, lies in rediscovering the wildness in us. Only when we fully appreciate the mysterious dimensions of our relationship to wildness will we begin to understand that it is one world inside and out; and that the preservation of nature is really nothing more or less than the preservation of the human spirit.

It was a gray, soggy morning, cold and wet like it had been for the last three days. The cedar and spruce stood dripping, hunched against the dark September sky, as I bent to the paddle. My mood just about matched that of the lake, and tattered veils of mist shrouded us both as I worked my way along the shoreline.

I was heading home from a four-day solo canoe trip. I stopped to pull down the brim of my old flop hat against the mizzling rain, then heard a familiar sound. Looking up into a spruce overhead, I saw a chickadee, and another and another . . . a dozen or so of the little bandits, flitting around the spruce, buzzing, hanging upside down, and generally acting ridiculous. "Psssh, psssh, psssh," I said, and they came down close.

I stared at them in fascination. For a moment it was as though I'd never *really* seen chickadees before. Sitting under that dripping spruce, I thought, "They must be crazy!" But I also vaguely wondered, "What ever happens if a chickadee gets depressed? Or doubts himself, or gives up?"

I sat there a long time, and I guess I decided that the fact there *are* so many chickadees, that they stay all through the winter, and that I was sitting there watching them in the freezing rain, kind of answered my question. They reminded me of the old woodsman's adage, "No matter how cold and wet you are, you're always warm and dry."

There have been many times since when I've thought about those chickadees, many times I've remembered them during a dark or difficult time—to the point that the chickadee has become a sort of personal symbol or talisman for me. But it wasn't until recently that I came across a Lakota Indian legend that crystallized my feeling.

It seems that in the long-ago times the evil powers were seeking to discourage and overcome humankind. They thought that perhaps through a long siege of cold and storms and famine they could reduce the race to despair. So, after creating these conditions they sent the little chickadee as a messenger—to find out the condition of the people, and to bring word back.

When the little bird arrived at the village he was treated with honor. He was fed and refreshed, and listened to closely as he told of his mission. When he finished, his hosts held a council, and formulated a reply. "Go back to the evil powers," he was told, "and tell them that the people are still living and hopeful, and ever will be; that they cannot be defeated by discouragement, storms, or stress, nor vanquished by hunger and hardships." And so, this is the message the chickadee brought back, and has been proclaiming ever since.

The Wisdom of Wildness

When I think about the chickadee, I think about the "wildness within," for he reminds me of the fascinating relationship between the human mind and the natural world, an idea Charles Lindbergh once termed "the wisdom of wildness." It is by no means a new idea. Longfellow once wrote, "All things are symbols: the external shows of Nature have their image in the mind." And nearly two thousand years ago Origen said, "Thou art a second world in miniature. The sun and the moon are within thee, and also the stars." I believe he was saying that human nature has been so deeply influenced by the conditions under which it developed that the mind is in some ways like a mirror of the cosmos.

Ultimately of course, the universe *is* wild. As William James wrote, it's "as game-flavored as a hawk's wing!" When we lie back on some rocky point in the Quetico and gaze at the night sky over pinnacles of spruce, what we are really seeing is wildness—endless, infinite wildness. In maintaining contact with wildness here on Earth, in preserving and nurturing wild places and things, we are in effect maintaining contact with the true nature of the universe, and with something deep within ourselves.

The Intangibles

One important element of this relationship is what my old friend Sigurd Olson used to call the "intangibles." Sig loved to talk about the intangible value of duck wings whistling over a sunrise marsh, or the first loon call of spring echoing down the sky. "Back of all concrete considerations," he said, "are always other factors which we call the intangibles. They give substance to the practical; they provide the reasons for everything we do. . . . If it was only a matter of saving representative bits of wildness, I would have given up my interest long ago, as would many others. Without the recognition that there is something deeper behind all this, there would have been no sustained effort to preserve nature anywhere."

So one of the first steps in understanding the wisdom of wildness is to know that it has much to do with intangible values, and has little to do with material values or exploitation. In the past decade, I've guided scores of wilderness trips, and have learned that what's *really* important to most people isn't how many fish they caught, or exactly how many miles they paddled, but the feelings, memories, and understandings they came away with—the development of a sense of meaning, of oneness with the land.

Intangible values are, of course, difficult to describe. That's why we call them intangible—hard to grab ahold of. Perhaps their meanings are best compared to those of a great work of art. Imagine trying to evaluate its worth in material terms alone— the size of the canvas, the amount of paint or clay or materials used. Certainly a great poem or a beautiful piece of music is far more than just the material value of ink spots on a piece of paper. Yet, we continually try to evaluate the natural world in just such terms. Even environmentalists, in trying to protect what little wildness and wilderness is left, sometimes use the same language. Inevitably such evaluations miss the essentials. We ignore the importance of the intangibles at our peril, for in doubting their relevance we trivialize the world. In materializing life we make it frivolous, and devoid of meaning.

Mystery

Another, deeper dimension of the wildness within is the awareness of mystery. Mystery is related to, and all tied up with, the intangibles, yet goes beyond them. In speaking of intangibles we are referring to an attitude of awareness, an appreciation that goes beyond the superficial; but when we speak of mystery we imply attitudes of awe, even reverence. Albert Einstein, a pretty fair physical scientist, once said, "The finest thing we can experience is the mysterious. It is the fundamental emotion which stands at the cradle of true art and true science. He who knows it not is as good as dead, a snuffed out candle."

As with the intangibles, we have difficulty in defining or explaining these deeper levels of mystery. Thus many Native Americans, when referring to the unseen meaning of power within all things, referred not to the Great Spirit as we've so often heard, but to the "Great Mystery."

Recently I guided a five-hundred-mile canoe trip in northern Saskatchewan. Along the way we became familiar with the Cree Indian legend of the Mammaygwessy. It is said these little six-fingered mischief makers torment the travelers of the far north, grabbing the gunwales of canoes in rapids, causing trouble around camp. For the remainder of our journey they bedeviled us, moving the kitchen utensils from one rock to another just when the cook knew exactly where they were, untying shoelaces and tripping people on the portage trails. We saw their pictures too, in ancient pictographs on overhanging rocks all along our route.

The Mammaygwessy are a part of the race of "dream people" known the world over. Their cousins are the gremlins, elves, and fairies, little people all, but large in the lore of humankind. We may laugh at such things now, but they are a part of our eternal groping toward mystery, the mystery that enfolds all of human life and experience, and to which one begins to feel close in the wilderness. So we paddled in the company of the Mammaygwessy, alert to mischief and the unexpected, and aware of mystery.

Throughout human history, people have been able to *touch* this dimension of mystery through just a few consistent methods—through myth and poetry and the arts; through ceremony which enables us to see the universe as sacred; and especially through personal subjective experience of the wild world. The modern paradox is this: the more we attempt to deal with reality as a material culture defines it—setting the world at arm's length, materializing and reducing it—the less *real* it becomes.

Today at the most advanced levels of modern science, nuclear physicists are saying the same thing. When you get deeply into the world of subatomic particles, you see your own mind in it. An electron is first of all your *concept* of the electron. Physicists tell us that the concept of the objective observer and the observed is obsolete, and if you hold on to it you can't get very far in nuclear science. It is not possible to separate the two because the very act of observing changes the phenomenon being observed. We are not observers, but participants; and this fact of participation means that the world is always Personal. Subjective. And Mysterious.

It may be that until we understand that life itself is rooted in mystery, we can understand very little else.

Resonance

But what has all of this got to do with the "wildness within" and our relationship to wild things? Simply this; the universe *is* wild, and there *is* a harmony, a resonance between wild nature and the human mind. We are made up of the same stuff as stars, mountains, redwood trees, eagles, and wood ticks. We are part and product of the same natural processes that produced them. In fact, when you really think about it, the very atoms of which we are made were once something else. We were once a rock, we were trees, now we are humans. Somehow there is memory, there is *resonance*. The natural world is inherently meaningful. It is symbolic and reflecting of the internal reality of human beings. Each of us is, in effect, a universe in miniature.

We *need* wildness, wild places and wild things, because they reflect to us who and what we are, and were, and may be. A world without wildness would be a room full of carnival house mirrors, all the reflections distorted. It would be like a locked studio in which the artist, surrounded and inspired only by his own creations, dissolves into irrelevance and madness.

We need wild things, psychologically and spiritually, because we see ourselves in them—and them in us. When we enjoy a fragrant water lily blooming on some solitary waterway, what are we admiring but the blossom that blooms in our own heart? When we throw a once-injured hawk into the sky, what are we celebrating but the freedom of our own spirit? When we marvel at the view from some high escarpment, what are we wondering at but the expanse of our own mind?

Such things remind us that it's one world inside and

out, and maybe that's what the "wildness within" is all about. As John Muir, grandfather of the modern environmental movement, wrote, "I went out for a walk, and finally concluded to stay out until sundown; for going out, I found, was really going in."

Harmony

So we have come to the idea of unity, unity between humankind and nature, between inner and outer worlds. But once again, there is a deeper level. The idea of unity has been abused for centuries, wherein someone's idea of unity is imposed on others through power. Much more important than unity, and underlying it, is a more subtle concept—called harmony. And harmony comes from within. You can picture the difference this way: you could take two cats, tie their tails together, and drape them over a clothesline. You'd have unity, but you'd be a long ways from harmony!

My friend Joseph Cornell once put it another way. "When you're out of harmony," he said, "you're kind of like an electric iron that's not plugged in. You can work real hard and do a lot of things, but you're really just pushing the wrinkles around."

Harmony is a beautiful concept, but how do we achieve it? Upon the answer to that question may depend not only the future of our wild area, but that of ourselves and our planet as a whole. The answer, I think, has a lot to do with the feeling of rain down the back of your neck. It has a great deal to do with sitting and pondering a sunset. It has to do with what the Menominee Indians meant when they said, "Look often at the stars and the moon." It has to do with coming as much as possible into direct physical contact with the primitive, and the elements of nature— sunshine, rain and snow, air and earth. It is concerned with times and places of solitude, of listening and awareness, and simply *being*. It means occasional separation from cultural artifice and media and support structures, and proximity to the cycles and rhythms of nature.

For thousands of years among indigenous and traditional peoples the world over, the goal and practice of harmony was the overarching priority of human life. It was understood instinctively that inner and outer worlds were symmetrical, and could not be separated. The idea can be symbolized beautifully by the bow and arrow. Simply put, before you can shoot an arrow well and accurately out into the world, you must first *pull it back to yourself.* Somewhere along the way we lost this understanding, and are now struggling to reclaim it.

Once I met a young Native American woman intent on reclaiming the wisdom of her people. She told me that one of the traditional teachings was that before you take any action of significance, you must first consider its meaning for the six preceding generations, and its consequences for the six following ones. I sat down for a long time after she said that.

The Problems

We have considered some interesting philosophical concepts, but do they really have any relevance to the incredible array of environmental challenges we now face? Let's look at some of the problems.

We now live in a world in which the "greenhouse effect," the accumulation of carbon dioxide in the atmosphere, is going to produce *extensive* climatic and environmental changes in the next few decades. We live in a world where industrial chlorofluorocarbons are destroying the ozone layer essential for life. We live in a time of unprecedented depletion and pollution of drinking water supplies, in which twenty-five million people a year die from water pollution; a world in which one-third of the earth's land surface is now classified as arid or semiarid, with thirty million acres a year being lost to soil erosion. It is estimated that within twelve years one-fifth of the remainder will be gone.

Some natural systems aren't just being abused or polluted, but lost completely. The United States, in its two centuries as a nation, has lost about 55 percent of the wetlands it started with, a priceless resource of biological diversity, water purification, and aquifer replenishment. The grassland prairie is nearly gone completely. In the tropics, rain forests are being destroyed at the rate of twenty-five acres a minute.

One result of this worldwide assault is that the global process of extinctions has also increased drastically. Paleontologists estimate that the rate of extinctions before the arrival of humans on earth averaged about one species per thousand years or so. Today that rate is closer to one species per hour and it is projected that the rate could increase to several hundred species per hour within our lifetime. This would translate into the extinction of half of all life forms now on earth by the year 2050.

But it is not the earth alone that is wounded. Humankind itself is no more healthy than the earth and her tattered natural systems. We know from the study of psychology that any *relationship* based solely upon manipulation and exploitation— one party manipulating the other to achieve his or her personal ends—is always unhealthy for both parties, and in the end is doomed. It seems we have been on the edge of a kind of madness, and are now trying to claw our way back from the brink.

Such societal problems as depression, aggression, and addiction are probably not understandable apart from the existential vacuum that underlies them. In other words, we have lost touch with our old friends the Intangibles, the underlying values and meanings that give life substance. In *materializing* our world, in systematically taking out all the meaning and merely exploiting it, we are left with—surprise!—a world of little meaning. We are left, ultimately, with poverty; a poverty of meaning, of beauty, a poverty of the spirit.

The Hope

I emphasized the word *relationship* above, for I believe that within that word lies the key to the wildness within, and our hope for life on this third sphere from the sun. Webster's defines relationship thus: "The state or character of being related or interrelated; working together—or being of the same kind; kinship."

Kinship. What a simple, extraordinary idea. What an essential idea. Among the Plains Indian peoples of America, no important ceremony was ever begun or completed without first the words, "All my relations." These words imply a powerful truth—that all living creatures—men, animals, plants—indeed all living systems and the natural elements themselves, are partners in the wonder and mystery of life; that other beings are our companions in this living world, relatives who should therefore be treated with respect.

I once heard Chief Oren Lyons say, "There is only one law, a law of light and dark; of blessing when followed, and suffering and destruction when not obeyed."

But it is not just the American Indian who has expressed this concept. Scientists like Albert Einstein, or René Dubos, who called for a "theology of the earth," Aldo Leopold in his magnificent statement of a "land ethic," all expressed the principle that we are a part of a community, a *family* of life—that we are literally related. The world over, throughout the varied cultures and spiritual traditions, the greatest thoughts of humanity have again and again pointed to the same insight:

From the Buddhist tradition—"One nature, perfect and pervading, circulates in all natures."

From the Hindu Bhagavad Gita—"In the light of understanding you will see the entire creation within your own soul."

From the Aborigines of Australia, who for thousands of years have said that human beings, animals, plants, and the earth itself are as one—all manifestations of a single life force.

From the Taoists of China, who say, "Do not ask whether the great Principle is in this or that; it is in all beings."

From the great twentieth-century Christian Albert Schweitzer, who based his entire life and work on the principle of "reverence for life."

And from the old monk in Dostoevsky's classic *The Brothers Karamozov*, who said, "Love all God's creations, both the whole and every grain of sand. Love every leaf, every ray of light. Love the animals, love the plants, love each separate thing. If you love each thing, you will perceive the divine Mystery in all, and you will come to love the whole world with a love that will then be all embracing, and universal."

As Chief Lyons said, there is indeed one great law, and that law is best described as a feeling or attitude written upon the heart—the attitude of reverence. When we act in accordance with it, we act in Harmony with the nature of life. When we fail to do so, we ultimately endanger ourselves and our world. We will not save the world with just our heads. To halt the destructive momentum and make a full turn, it will take a change of heart, followed by a change of mind.

There truly is a "wildness within," and it is time—past time—to rediscover it, to honor it, and to honor "all our relations." As Sigurd Olson put it thirty years ago, "The preservation of waters, forests, soils, and wildlife are all involved with the preservation of the human spirit."

The chickadee would undoubtably agree, and would expect no less of us.

First Ice
Dancing on Water, with a World at Your Feet

Peter M. Leschak

Mark Lake was a sheet of diamond. During a dead-still night in late November, a polar air mass had crept down from Manitoba and settled in to change the world. The limpid, mirrorlike surface of the lake had become a mirror in fact. The reflected images of stars were slowly encased in crystal, and by dawn the winter constellations were glinting off a one-inch mantle of ice.

The weather stayed clear and cold, and now, two days later, we were tentatively exploring the integrity of the sheet. We had never seen such hard ice. Unless your skates were freshly honed, capable of biting, it was impossible to stand up. We started at the shore with a cordless electric drill, boring a series of test holes farther and farther out, until we were satisfied that the lake held three inches of solid ice clear out to the middle.

A lake that freezes on a calm night is deceptive. The ice is as transparent as a window, with few bubbles, and it's hazardous to gauge its thickness by just looking. If there's a crack, it usually extends from top to bottom—and you can study the seam to make a guess. But refraction and depth perception are tricky. So we trust the drill. The drill has never failed us.

Even so, that first sweeping glide away from the land is like a stolen kiss, or the ascent up a tall, shaky ladder. It sends a fluttering charge down the spine and out to the fingertips. It's an act of faith that you can walk on water. Most skaters start out slowly, involuntarily hunching up their shoulders to make themselves "lighter." But in just a few moments, the fear vaporizes in a rapidly building sensation of flight—for on clear ice, there is a world beneath your skates, a world of hills, and valleys, and forests of aquatic plants. As you glide over shallow water, the lake bottom races by, as vivid as a crisply focused color photograph pressed under glass. There are no waves to scatter sunlight, no waves to swirl sediment into clouds. On Mark Lake, I imagined that I was orbiting an alien planet, zooming above an atmosphere of ice. I could see the submarine world as plainly as if I were in it. But I was not in it. I was soaring through my own open space, out in the freedom of blue air.

Soon we were whooping and shouting, pumping along the shoreline, cutting in and out of the little bays, skirting the frosty trunks of half-sunken deadfalls. Three months earlier we'd drifted beside these same logs in a canoe, lazily casting for largemouth bass. But now all was speed and sailing.

We were intoxicated—overcome by the joy of pure motion. It was as if the regimes of friction and gravity had fallen with the temperature. How little effort it took to whip down the lake! This must be how an osprey feels when it catches a thermal over open water and floats on swells of air. We were as free as you can get in two dimensions, unbridled on our flat and vitreous reflection of the sky.

At the east end of the lake, in a shallow nook that's good for bass and frogs, there was a wide mosaic of green lily pads, frozen into the ice. It was a magnificent still life: last August's verdure cast in icy plexiglass. As I flew over, a movement caught my eye. I skidded to a halt in a shower of crystalline dust. Something was squirming under the ice. I dropped to hands and knees and peered through. It was a huge water beetle, crawling along the bottom of the sheet. Its body was about four inches long and almost two inches wide; its black, spindly legs seemed out of place underwater. It was upside down, belly toward the surface, and as I slid forward, following, we were face to face. I knocked on the ice with my fist, but there was no reaction—no stopping, no change of pace. We were on opposite sides of a thin but profound barrier—in separate universes. I was only three inches away, but the beetle had no inkling of my existence. I was like an omniscient but impotent god—able to survey the cosmos, but unable to enter it.

Mark Lake is locked up for the winter, the ice tempered and polished like some strange, gleaming alloy. We are exhilarated and rich; we own fifteen acres of smooth ice. Darting away from the shoreline, we rip out across the lake, digging hard and swinging our arms like speed skaters. We run out of energy before running out of ice, then rock back and just glide, or twirl into figure eights and yell at the sky. A few years ago, we saw fish below the ice and spent several crazy minutes "herding" a school of shining perch before they dived for deep water. Flashing skates on a frozen lake are a tonic that restores youth.

In *Walden*, Thoreau wrote: "The first ice is especially interesting and perfect, being hard, dark, and transparent." Through its perfection, we are initiated into the wizardry of motion. We learn how to dance on water.

THE ORDER OF THINGS

Paul Lehmberg

Earth, air, fire, water. Empedocles taught us 2,500 years ago that they are the stuff of the earth, the "roots of all." I believe him. In the Quetico it is easy to discount the chemist's periodic chart for its slavish literalness, and though Empedocles' cosmology is commonly thought to be outdated, I am partial to his schema because it is good poetry and transcends obfuscation. It is true that his world view wanders from the letter of things, but its simplicity mirrors the simplicity that undergirds this land, and his cosmology reflects the spirit of this country. Besides, a geography as lacking in dissemblance as this one deserves to be composed of 4 elements, not 105.

Because so little is hidden here, the Quetico is an excellent place to begin a study of geology. The most amateur of geologists can sleuth successfully here, since it takes neither the eye of a trained expert nor the imagination of a fevered poet to marry the explanations and descriptions in the guidebooks with the realities of the landscape. Basic knowledge is easily won, and the place makes sense because appearances do not deceive. A visitor here, wary of appearances, once looked down the steep scribbled shoreline and remarked offhandedly how Nym looks just like a hole filled with water. She was from southern Minnesota where sandy-shored lakes rest on the land like drops of water on a glass slide, and it took her a moment to realize how ridiculously right her passing comment had been. It is freakish to be granted such free access to reality, and she laughed at the absurd delight in having made such an obvious rediscovery.

It is the recent advance and retreat of a series of glaciers that have made the land as ingenuous as this. During the last geological epoch (the Pleistocene—2,500,000 to 10,000 years ago), Earth's climate cooled several degrees and snows in the region of what is now Hudson's Bay no longer melted. Piling up layer upon layer, the mass of snow became ice, and when it reached a thickness of three hundred feet the body of snow and ice began to move. It had become a glacier. Spreading out like batter poured onto a griddle, it lumbered south over the Canadian Shield and into the Quetico at the rate of an inch to ten feet a day. By cementing uptorn rock debris into its snout and underbelly as it heaved down out of the North, the glacier quarried for itself, cutting, scraping, gouging edges and planes that completely destroyed existing drainage patterns in its track, and the sheer weight of one of these icy leviathans, some of which grew to a thickness of two miles, flattened the crust of the earth. Nothing could stop the slow cataclysmic advance but a body of seawater or a rise in the mean temperature.

The fourth and last (but for the next) glacier met its end as a result of a climatic change that occurred 15,000 years ago. The earth began to warm again and the glacier, shriveling in the sun, began to retreat—or die—northward, losing its mass and leaving behind a barren profusion of scoured and tortured bedrock littered with gravel, rocks, and boulders, all awash in the meltwaters of rotting hulks of ice stranded like whales on a beach. The glaciers had stripped away intervening Earth history and left exposed a moonscape of Precambrian rock—rock of the same age as that buried at the bottom of the Grand Canyon.

Geologically speaking, the last advance of ice is today's front-page news. If, to add some meaning to the fantastic numbers involved, we telescope Earth's history into one year—now is the stroke of midnight on New Year's Eve—then the last glacier began its retreat less than fifteen minutes ago. Since this glacier made its exit only a quarter of an hour before our appearance, it is, of course, no surprise to find fresh tracks that are easy to read. The striated exposed rock surfaces were caused by the brush of rock against rock as the glacier moved south. The northeast-southwest grain of these elongated lakes, so obvious on a map of the Quetico, is the track left by the glacier as it furrowed its way across the land. The thin and ragged cover of soil is as telling as fresh paint that this is newly scoured land living out the first minutes of its latest incarnation. Even those strange spooky boulders at odd spots in the woods are attributable to the glaciers. Some of them are as big as three-story houses, and they wait almost expectantly, as if eons ago they were left behind for some obscure reason by traveling giants who will return for them in the next few days, hopefully not while you are around pondering one of them. The boulders are called erratics, lost out of the glacier as it made its melting retreat. This land is newly re-formed and undisguised, and if you cannot infer from its features that it was formed by a glacier, you need only read the word *glacier* or have it whispered in your ear, and the oddities and questions explain themselves.

If one chooses to trace local Earth history past the Pleistocene and back toward Earth's beginnings, there are, to be sure, obscurities enough to stump even professional geologists. Yet they, just as much as fair-weather geologists, are attracted to the Canadian Shield, and for much the same reason, though professionals study beginnings here instead of endings. The Canadian Shield is unique among the great continental shields because of its vast expanses of exposed Precambrian rock, which by definition was formed after Earth cooled and before organic life had evolved to the point where it began leaving fossil traces. On the Shield geologists have access to Earth primeval, and they hope their study of these ancient rocks will fill in the missing chips in the mosaic of how Earth was formed. Yet, whatever professional geologists eventually extract from their studies of the Shield, the events of the latest geological age are clear, and when we happen on simplicity it behooves us not to commit the unnatural act of burying it, and maybe ourselves, in a honeyhead of our own making. In the Quetico you can drop your defense against mere appearance. Each lake is the quintessential lake, like a Platonic form; whatever else a lake may be, whether it be spring- or rain- or snow- or stream-fed, it is at the least a hole filled with water.

The prevailing geologic mood is simplicity, and because we wish to fit here, we strive to maintain and emulate that simplicity. Clutter, either the appearance or the reality of it, is anathema here, and its nettlesome presence is to be watched for and swept away like a cobweb lest it mutate into chains. Therefore the cabin we would build had itself to be simple so that it would fit here. The obvious plan would have been to build a log cabin out of our own timber, but our land patent, which was only provisional until we had an enclosed structure, did not include the timber rights. But our lack of building experience would also have prevented us from building with logs. Having built nothing but the usual birdhouses and coffee tables in high school shop classes, we knew we were no more than wood butchers, and that it would be unwise to tempt providence by trying to build our cabin of anything but sized and finished lumber. But if not a log cabin, what?

If we were almost totally lacking experience in the building arts, then maybe it would make sense to clear our cabin site and then air-ferry in a factory-made metal trailer complete with carpeting and a brightly colored conical fireplace like the ones in the catalogues, or if not that, then to hire a carpenter who could help us with the design and who knew what he was about when it came to carpentry. No one even bothered, fortunately, to put forth the first suggestion, all of us agreeing without discussion that such gaucheness combines the drabness of a boxcar with the visual affront of an empty beer can as big as a boxcar. Aside from that, a man should be responsible for building his own house just as he is for washing his own body (and if idealism flags here, is there a chance for it anywhere?), a belief that also did away with the latter possibility.

We considered ordering from one of the lumberyards in Atikokan a prefabricated cabin, its pieces already cut out and numbered and ready to put together like a set of Tinkertoys or

Lincoln Logs. There were objections to that, too. We wanted a cabin with our imprint on the design, and not that of some anonymous architect in Minneapolis or Miami who knew nothing of our wishes or needs; further, we wanted a cabin that was made to use and not only to sell. We would, we decided, design it ourselves and construct it ourselves, and if the product turned out to be more like a cartoon of a cabin, or if through inexpert workmanship or faulty design it collapsed the following winter, so be it.

How about an A-frame, then, one of our design and construction? someone asked. No. Its shape, though simple, is inefficient, and its simplicity is the kind that draws too much attention to itself. We were not here to out-Jones the Joneses by one-upping our neighbors the trees. A split-level, then, sided with plywood, something of modest design, but unique, tailormade for a special spot on the hill? Too ambitious, we knew, and we had come full circle, reminding ourselves again that we were anything but artisans.

During the months we spent considering various cabin plans through letters sent back and forth from Minnesota, Illinois, and Utah, Rick described for us a modernistic chalet he had seen once in the Swiss Alps. Three stories high and large enough to be a hotel, it was instead a private home. The chalet was rocked and timbered into the mountainside, and it clung there proudly— and, perhaps, precariously. Two sides of the living room fronted the valley, and massive, widely spaced beams were all that separated the main floor in the living room from the roof, which was three stories and thirty feet above. Windows on the valley sides of the chalet stretched the full height of the living room and offered an unobstructed and—need it be said?—sublime view of crag- and ice-encrusted Alps. Clearly the stuff of which picture postcards are made, I thought as I read, but what did this have to do with Northwoods cabins?

Rick suggested the outlandish, that this chalet, this modern-day castle, had just the sort of floorplan we wanted. Once it was stripped of its rich furnishings, its adornments, repetitions, and excesses, there was the basis of a building plan here that was simple and serviceable, and one that three apprentices could construct without benefit of a foreman. Rick roughed out the basic floorplan, and as we happily discovered in the next exchange of letters, we were all excited by the simplicity and efficiency the plan promised. In later correspondence and meetings we traded suggestions for changes and refinements, and after our consultations with those who had more experience than we, the sketches graduated to drawings and then to blueprints, which were neither blue nor orthodox but, for all that, still exact and readable by anyone. The final step in the preliminaries was to construct a balsa-wood scale model that included every detail except for the furnishings.

But we did not build the cabin first. We began with our outhouse, a two-holer (to spread the pile), which we built a couple of feet longer than necessary so that we could use part of it as a tool shed. We began with the outhouse first, not because we especially crave the creature comforts, but so that we might acquire experience where mistakes would not count so much. Our only error, committed late one afternoon while we were hurrying to finish measuring, cutting, and nailing on the siding, was to saw in two a section of the siding we had been saving for the express purpose of serving as our door. Sheepishly, we acknowledged what everyone is already supposed to know, that haste makes waste. In spite of our error, which turned out not to be so wasteful after all, since we were able to salvage the pieces for our door, when we finished we had an outhouse that was plumb and square, and what is more, the confidence that the following summer we could carpenter a cabin equally well built.

As planned, we returned the following summer and in two weeks of long days built the cabin. The scale model, which is just a toy now sitting on top of the bookshelf, assisted us in seeing how the parts were related to the whole, and we consulted it whenever our minds balked at thinking in three dimensions. Each night when darkness, or Suzanne's and Janet's call for dinner, forced us to take off our carpenter's aprons and stow our tools under the foundation, we would walk halfway down the path toward the kitchen and then turn to survey what we had built that day, and each time we were not a little surprised that we were the carpenters—yes, carpenters!—on that cabin which was so quickly going up. Again we discovered that patience, thoughtfulness, and planning can serve as effective substitutes for experience. A few stupidities committed in the act of construction resulted in some minor though negligible errors, and they were either removed or else expertly hidden. The only structural flaw that I have been able to detect—and that with the help of a plumb bob dangled from the roof peak—is that the roof leans a little to the northwest, but the list is so small that it is doubtful the laws of construction will exact any retribution for a good long while. Though there are some minor mistakes, the cabin is as simple and sturdy as we wished it would be, and we admit to being a little proud of our humble cabin.

The owner of the Swiss chalet, and probably his architect along with him, might regard their progeny as little more than a bastard child, and when we want to put on some false airs for the fun of it, we remind each other of the monied, high-toned lineage into which we arranged for our cabin to be born. The chalet reaches toward grandeur; our cabin would never be accused of that ambition, though perhaps of others. But, bastard or not, the cabin presents a face that more than her builders can love. It is well and lovingly made of high-quality, kiln-dried Canadian lumber. It is stout and steep-roofed and as strong as a seaman's chest. A perfect square twenty feet on a side, its cornerstone an outcropping of the Shield itself, the cabin rests on a foundation of eight additional cement pads topped with anywhere from one to four cinderblocks, according to what the terrain demands. The cabin is framed with pine and sheathed with aspenite, which is a particleboard pressed from aspen chips that is much used for building in this country, and it is capped with a traditional gable roof. Despite its plain, square, somewhat stubby look, we felt no need to jive it up by painting the exterior fire-engine red or banana yellow. We are partial to the color scheme that was in evidence when we arrived, and so we stained the cabin with a wood preservative that matches the rich, dark brown of wet jackpine bark, and we shingled the roof in forest green, except for a few strips at the peak where we ran out of forest green and had to finish with a lighter, dusty-green shingle.

The superstructure of the cabin is as ingenuous as the rock on which it rests. The skeleton is freely visible, and nothing is obscured or hidden by trim or finishing work. You can see how the frame of the cabin *works* to adjust and balance force and weight. Spacers have taken out the natural bows and small warps in the wall studs, thus perfecting their vertical rise and improving their parallel run, which in turn squares and strengthens the frame. Sheets of aspenite that cover the frame freeze its square lines into place. Horizontal ceiling joists, which span the cabin and are spiked into the rafters near the eaves, balance the outward thrust of the rafters, which at the roof peak lean into the ridgeplate. Vertical kingposts, which stretch from joist to rafter and form part of the roof trusses, relieve the rafters of some of the roof weight by distributing it onto the joists.

On the floorplan, identical sleeping alcoves—it would be pretentious to call them bedrooms—large enough for only a double bed and a small dresser sit at diagonal corners. At the other diagonal—but words are obscuring the lines:

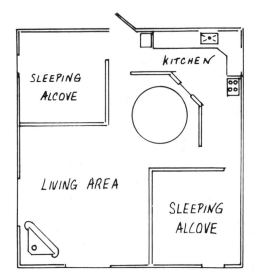

MAIN FLOORPLAN

Like the living room in the chalet, our living area is open to the roof, except for two six-by-ten-inch beams that traverse it, and if our ceiling were not writ so small, we too would have what the developers call, whether it is writ small or not, a cathedral ceiling.

A ladder, which pulls out from storage against the northeast alcove wall, leads through a trapdoor above the dining area to the loft:

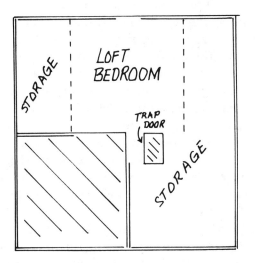

LOFT FLOORPLAN

A waist-high railing borders the loft overlooking the living area, and because the loft is not walled in at the kingposts, there is a pronounced spaciousness to it despite its modest dimensions. Bordering the loft behind the kingposts is ample storage—much more, in fact, than we are using.

There is an aesthetic to the design of this cabin, and the aesthetic is based partly on symmetry. It is a shining mystery why one's sense of the beautiful should resonate to the spatial rhythm that is called symmetry. We usually hold dear the rare rather than the commonplace, and spatial and aural rhythms are commonplace. Life begins when sperm and egg fuse after the rhythm of procreation; it ends when the rhythms of life breath and heartbeat are broken; the interim is spent in a body outwardly symmetrical. Our lives are bound into rhythms as if they were the covers of a book and we were the pages, and it is a felicitous state of affairs, to say the least, that we hearken to what there is so much of and that we seek even to create our own play of rhythms.

Rhythm is commonplace but seldom common, and thus it is that I am satisfied with—and by—the simple repetitions of the cabin: the two sleeping alcoves, identical in size and mirror images of each other down to their dropleaf windows; the two bay windows, each centered in a living-room wall and equidistant from the focal point of the room, which is the Franklin stove sitting in the corner; the two kitchen windows, one an echo of the other, equidistant from the corner; the loft bedroom centered in the cabin between the two rows of six kingposts; and the loft window centered beneath the apex of the roof. Yet the symmetry of the cabin is not complete. There are grace lines that save the cabin from a honeycomb tropism: the division of the cabin into fourths that are only approximate; the unbroken line of sight on only the one diagonal—from the kitchen, through the serving window in the dining area partition, to the living area; the open-ceilinged living room; the total dissimilarity between the loft and the main floor. Such contrapuntal lines as these rescue the design from the cold tedium of the hive.

We chose the design not only because it was simple, but because we saw in it the promise of function. The American sculptor and aesthetician Horatio Greenough equated the promise of function with beauty, and we tacitly accepted that part of Greenough's theory which states that an object is beautiful in proportion to how well it works. The form had to be adapted to the function, and each of our rooms is planned around its use. The sleeping alcoves might be considered tiny, even cramped, but they are expansive enough for two people to close their eyes and spend eight oblivious and restful hours. The small dropleaf windows, about the size of a common basement window, provide each alcove with good ventilation and light.

The kitchen is likewise designed around its use. It is not inefficiently strung out along one wall but is L-shaped, so that it wraps around the cook to cut inefficient and unnecessary movement. The windows are near the most often used work areas, the stove and sink, and there is plenty of counterspace—more, in fact, than I have seen in many apartments. Underneath the counter are rough shelves where we keep our dishes in full view and within easy reach, and below them are our staples, all clearly marked in coffee cans or canisters. The kitchen side of the dining-area partition is shelved with scraps of wood, which provides us with additional storage.

Since the living area is where we spend most of our time, it is the largest room, of course, and the one where we most wanted to make efficient use of our limited space. Its open ceiling and the open line of sight into the kitchen combine to expand the room psychologically. The two large windows, which are screened at the bottom, provide sufficient ventilation and excellent light, even on cloudy days, and through each of them we have a fine view of the lake.

Except in the kitchen, the furnishings are as spare as the plan is uncomplicated. Strangely, we have three or four incomplete sets of dishes, sauce pans and stew pots and frying pans of every size, coffee pots for two and for twenty and for every size group in between. We could not serve an army—but probably a small church supper (and I suspect that many of these vessels have). But, aside from the surprising variety and number of dishes, we have: a bed and dresser, a mattress in the loft, a chair and couch, a treadle sewing machine housed in a wood cabinet, a small chair-side table, a kitchen table and chairs, and in the kitchen an icebox—the kind you put ice in—and a propane stove. And not much else.

It is because of a cloudburst yesterday afternoon that I have been thinking of the grooved lakes and the ragged carpet of fauna that has recently been rolled down here and the conscious simplicity we have tried not to lay on but to build into Nym. All day there was that nervous stillness that presages a storm, and an almost palpable mass of stagnant air lay over the land. Then in midafternoon there was the coming of the winds and after that the deluge itself, the sodden clouds splitting and spilling their torrent of water on the earth. From the window I watched, trying to part the curtain of water that hung from the eaves, and through it I caught glimpses of something I had never seen here before, and I was as startled as if I had seen a man suddenly appear from nowhere at the window. There was a stream of water running in my midst! The path to the Toe had become a rivulet of muddy water with a feisty current, and I traced its course down the hill and out of sight. Along its edges the wash of water had in a matter of only fifteen minutes formed alluvial fans and hillocks of pine needles, mud, and wood chips. "Je-sus," I swore under my breath, not because I had thought the country erosion-free, but because thoughtfulness and good intentions were not enough. While I stood there washed in the afterwaves of my own astonishment, I flashed into futures for Nym that were all full of the wrong possibilities and all of them my own doing—neon and concrete and steel and plastic and all things synthetic, dustbowls and amusement parks and ore trucks corkscrewing their way up out of a drained Nym Lake just as they are doing now out of Steep Rock Lake not twenty miles away. The running sore out the window was at once an insult and an embarrassment and I was both resentful of and apologetic for it.

Can't I even walk on the land without gutting it? Maybe not: Acts have their consequences. If it does happen that without being aware of it in time, we immutably change or destroy what we have sought and found, we will, of course, not be unique. On the contrary. In Western culture that tradition dates back to Adam and Eve, who can be thought of as the patron saints of the destructive arts. Even in the Quetico we would not be unique, the fur traders and lumbermen having antedated us by a good many years. North Americans have demonstrated a miserably small capacity to husband their Eden. In fact, we have even had an ill time of it nurturing what we ourselves have built.

Rick, Bob, and I have some knowledge of the quiet paths that such destruction can take. That we built Nym is due largely to the fact that while we were in high school the three of us, along with nine others, built what in that Minnesota locale was called a hunting shack, which meant a sparsely furnished cabin not far different in conception from what I am living in now. Having decided that life was more worth living if you had a shack where you could hunt and try out your incipient adulthood with the guiding hand of parents and society resting a little more lightly than usual, we convinced one of our fathers to give us two rundown Quonset huts that had at one time housed turkeys, and we transported them one at a time on a flatbed trailer to a patch of soilbank that bordered a tree-lined pothole about ten miles from town. We had acquired squatters' rights there by politely asking the owner for them.

We tore out two ends and butted the Quonsets together on a foundation of railroad ties, shoveled out the turkey droppings and disinfected the building with the strongest chemicals we could find, tar-papered the roof, replaced a few window panes and rotting boards, painted the inside, and furnished it with a barrel stove and a table and some chairs. In a month we had our shack, and it was a simple, spare, and exciting place.

But we couldn't sit still. Pascal claimed that all man's troubles stem from his inability to sit still. That may not be an oversimplification. In our case, at least, he was right, literally right. With access to twelve basements and attics, we grew restless—we had discovered what a cornucopia of necessities each of them was, and we began to pick them over in search of items that would add further improvements to the shack. Various tools and cans of paint began to appear, and soon we had to partition off a tool room at one end of the shack in order to neatly house our excesses. It was discovered that we had enough baby-blue paint to cover the whole interior, so, of course, we were obligated to re-paint. When we were done, everyone agreed that the new color was vastly superior to the old green, which was a vile shade that had looked old even when it was still wet—we had come by the green through chance, having had to mix all our paints together to get enough of the same color for one complete coat. Piece by piece additional furnishings began to appear, a medicine chest, a kitchen sink, a rollaway, some easy chairs, a double bed, an end table or two, even a battery-powered radio. Someone scrounged some carpeting, and what of the new flooring we couldn't use we stored under the shack, along with the overflow from the tool room. And so it went.

The winter of our freshman year in college we burned it to the ground. Our public reason for the burning was that, the group having split up and left home, the shack was ripe for usurpation by another set of squatters, underclassmen, whom we regarded as unworthy barbarians because they had not toiled as we had. The private and unspoken reason was that the burning signified our acceptance of the end of something, and what we burned on that bitterly cold Sunday afternoon was what many of us, I think, had ceased to want, and certainly what none of us had planned or foreseen three years before. The dirt and clutter had gotten the best of us, gotten away from us, and we had grown claustrophobic from all the clutter, which by imperceptible turns had grown more and more shabby and which in the end was only posing as comfort. In the alternating humidity and dust of the midwestern summer, the overstuffed furniture in the overstuffed shack had gone ratty with mildew. The rug zealously collected dust as if it had been laid for that very purpose, and the threadbare chair and mattress coverings clung to the skin like the sticky hot and vaguely verminous sheets off a sickbed. So it was with a measure of relief that we burned it, and afterward I regretted that one of the most noxious comforts could not disintegrate with the rest of them in that explosion of fire. Through the smoke near the charred railroad ties, I could still see a raw mound of dirt, the effluent from a landscaped patio that someone had begun to gouge from the hillside.

We met defeat once at the hands of restlessness, but now we have some idea of the nature of that sly continuum of acts that like a stagnant river appears innocent of movement, but that in time reaches the sea just the same. To single out this or that change as the most decisive or most important—acquiring our third hammer, for example, or laying the carpet—is as fruitless as trying to separate out the most important ten gallons of water in the river. Even assuming there is a decisive act, it is probable that our dull minds would miss it. It is like watching the hour hand on a wristwatch; we miss the movement because it is too fine and we see only the results. When you have come so far from the idea of a shack that you are considering wall-to-wall carpeting, the general course is clear even if its particulars are not, and the truly decisive moment—or, more likely, moments—is in that hazy, imprecisely plotted time when carpeting became a live possibility. Decisions are not made in a vacuum: They are made in the ambience created by the accumulation of past decisions. It pays to think of each act in the continuum as a vector that by definition has both direction and, what is most important—but only because we forget this so easily—momentum. Enough acts committed in the direction of complexity, and you are not only pointed in the wrong direction, you also have less chance, because of the combined momentum of events, of redirecting yourself.

It is hard to say at what point you ought to make your stand against the onslaught of clutter and complexity, but before your direction and momentum are glaringly obvious, and certainly before the innocent-minded willingly aid in the defense. If I am right, you must make an arbitrary and capricious judgment like the one I wish in retrospect I had made: No, it is absolutely crucial that we have only two hammers, not three. Such a pronouncement would have inspired all sorts of verbal abuse, and I would have been damned as ornery, arbitrary, a nitpicker, and—no doubt—a fool.

And my friends would have been right—it would have been a preposterous demand, and I would not argue that acquiring that third hammer was a watershed event. What I would argue, however, is that we could have suffered healthily under this delusion could we have been somehow convinced into believing it. Clutter grows like fire and it should be fought like fire. The practiced firefighter plows his firebreak at a safe, "arbitrary" point downwind of the fire to ensure that it will be contained. It is quibbling to argue that the firebreak be plowed exactly in the spot it is; it is not quibbling—in fact, it is crucial—to argue that a firebreak be cleared somewhere in that vicinity. The hammer is in the right vicinity, and so it is that I make a symbol of it and suppose that it had in it all the evils that were in reality spread through the continuum.

We now guard against the recurrence of the past and are more careful than we otherwise might be to sit still, and we have a certain trust and confidence in ourselves. We want, we say, to keep Nym simple. But I find myself wondering how far our trust in ourselves should reasonably extend. I see signs that we have begun to stir from our restfulness. I have said that the cabin is a wooden tent and that its lines are free, but that is not quite true, at least not any longer, not since we insulated and then, to cover the dirty-pink insulation, paneled the walls with a type of particleboard often used for subflooring. We decided to winterize the cabin for the obvious reason that to do so would greatly extend its usefulness in this subarctic climate. Without insulation the Franklin stove did keep the cabin at a temperature hovering somewhere between the bearable and comfortable marks, but the certitude of comfort that insulation provides will encourage us to use Nym more in the winter. It is also true that in deciding to insulate we had our collective eye turned toward the future—that part of it just beyond what is called the foreseeable future—when each of us hopes to realize the wish to winter here. If we grant ourselves the wish, a winterized cabin would be, if not a necessity, indistinguishable from it.

We have increased the utility of Nym by improving it, and to my surprise the addition of insulation and paneling has not diminished the uncommonly large psychological space that we have managed to fit into these modest dimensions. My surprise is welcome, but at the same time it gives me pause to question how accurately we can predict the effects of future improvements. In this democratically run cabin, the decision to insulate was unanimous—enthusiastically so, in fact—but I wonder how controlled the step really was. The framing is now hidden by the paneling, and I miss—more than I thought I would—the strict geometrics of the wall studs bisected by the spacers, and I miss the feeling of being not much inside—or inside by only a three-eighths-inch thickness of aspenite. Because of a shortage of insulation we have not yet insulated the roof, so its lines are still free, but I wonder whether when the rafters are closed off, our improvement will not muddy to oblivion the music of rain beating on the roof. But if it does, maybe I won't miss the music.

There are other signs of our refusal, or maybe our inability, to sit still for very long. Gone are the days when against my will I tried to decipher the cranky metaphysics of a succession of fitful and failing outboards. A couple years ago we invested in a fine, new ten-horse outboard, which thus far has been exactly what we wanted it to be, the *ne plus ultra* of dependability. There is now talk of replacing our battered boat with a new and deeper one, the argument being that our present one is not so much old as dangerously shallow for a lake the size of Nym. To have to rely on this boat for transportation, especially in autumn winds so stiff and steady they seem machine-made, is to be imbued with the forcefulness and essential rightness of that argument. But if we purchase a boat, we also ought to replace the primitive boat slip at the Toe with a more permanent dock. I have noticed that on some days the boat edges toward the Toe and ever so lightly brushes against it, and some of the rivets near the transom are being ground away. We would certainly not want to subject a new boat to similar treatment, and the construction of a dock, properly lined with old car tires, would not be a luxury so much as a simple act of preventive maintenance. The dock, however, would advertise our presence here, which would defeat our having set the cabin back in the woods as insurance against being rifled by latter-day Vandals, who pillage even around here these days. At one time we considered adding a porch, but we discovered that the cabin is sufficiently cross-ventilated. Having given up on the porch, we now think of building a sauna, which would, admittedly, be a summer luxury—but it would also be a winter necessity. I have even picked out a tentative site bordering the clearing behind the Toe, and I know just how the runway of two-by-ten-inch planks will extend from the sauna door to the Toe and the restorative icy waters beyond. Thinking about it, I can almost feel the assaulting waves of dry heat from the water-spattered red-hot rocks and that delicious plunge into the water that purges body and mind of sundry discomforts. It will be hard, I know, to get along without that sauna.

As it will be hard to get along without the boat and the dock and the succession of other necessities that, apparently, will from time to time come looking for us to present their wares for our consideration. Such improvements are like people, individually agreeable, but who in the aggregate sometimes turn uncontrollable. Though many of these necessities will prove quite attractive, it remains to be seen whether we will succumb to their individual blandishments. But even now—fortunately now—it is a tiresome thought to consider the stewardship involved in keeping track of and maintaining an increase in wealth and complexity. We do not want to be stewards of a concoction of clutter, and improvements, if they are not improvements enough, will point us in a direction and spawn a momentum we never intended. So we are wary, but in a healthy and not a pathological way, of our direction and speed, and we choose to go, if at all, slowly.

It may be, of course, that we have no ultimately efficacious control over the character which this cabin assumes. Civilizations are said by some to have a life span in the same way that a sexually reproducing plant or animal does. Death is a built-in fact of life, and it may be the same with our cabin. I look up at the roof and it suggests just that. It is steep-pitched so that its lines will relieve it of the weight of winter snow and so that we could include a loft. But there is another reason for that steep pitch. If we ever need extra storage or sleeping space, we can, with a minimum of trouble, simply raise the roof and add a dormer. When I consider this last option that we have reserved for ourselves, I wonder if we haven't designed into the cabin the obsolescence of its simplicity. It is as if we have subconsciously anticipated the certain appearance of clutter, which is the prodigal that without fail returns and for which we have prepared a place, not because we love it, but because we recognize it as family.

Maybe it is illusion that our experience with the shack will ever be of any real benefit. But until convinced otherwise, I have to believe that experience is worth more than to forestall the inevitable. Yet, even though I cast my lot with experience, I would not want to rely on it exclusively to save us from our own tinkering. Fortunately, there is help proffered from other quarters.

Sloth, traditionally considered one of the seven deadly sins, may not prove so deadly here. In fact, she may be our guardian angel. At least she is working for me now. There are a number of small improvements that I have scheduled for myself this summer, and one of them is to hook up a trap and drainage line to the kitchen sink so that we will be spared the inconvenience of having to dump a potful of slops after doing the dishes. Even though it has been four years since Rick and I dug the outhouse hole, I have had enough of digging for a while yet, and the task of hooking up the line and burying it remains undone. I just haven't gotten around to it. And I don't really expect that I will unless boredom or a new wave of Puritan guilt gets the best of me some afternoon.

I have my reasons. Sinks are generally thought to be unworkable—worthless—unless they are fitted out with a drainpipe, but because of our hurry to use the sink once it was permanently installed in the kitchen counter, we temporarily placed under it one of those huge pots that had once served in a church kitchen, and we were quickly disabused of the error that a sink requires a drainage line to be useful. Our sink is a marvel of simplicity. When I pull the plug the water swirls down into the pot, and then I carry it outdoors and dump it in a natural depression about fifty feet from the cabin. There is no pipe to drain or to take in each fall and put out each spring; there is no pipe to clog and no plunger or plumber's snake necessary to unclog it. The drainless sink is error-free and bother-free, and I for one am willing to carry out the slops to maintain the best of all possible worlds, which in this case is the convenience of a large and permanent sink combined with the positive luxury of knowing that the sink will never be the source of any irritation whatever. A sink with a drainage line is not only not a necessity; it is a bogus convenience I can do handsomely without.

But having said that, I myself am not entirely convinced by this justification for laziness. That is, I believe the argument, but only a little—probably because it is a bit shrill and tries too hard. But if I were persuaded by my own argument any more than I am, my laziness would lose its purity, and, in any event, I know that procrastination and my disinclination to putter, and not this rationalization for my nonaction, are carrying the day and keeping me from adding one more worthless convenience.

My tongue is in my cheek? Maybe a little, but only insofar as I am contrary enough to enjoy being saved by a deadly sin like sloth. That personal quirk, however, should not obscure the fact that a little more mindless laziness mixed with all that mindless activity in our high school days would have at least checked the clutter which eventually encrusted the place, and sloth can do the same for us here.

Supposing that every other defense fails to save us from ourselves, there is one last defense, and that is Nym's isolation. It may be the best defense of all. Having been required to build a cabin here to obtain our land patent, we are homesteaders of a sort, and we like to think of ourselves as such, but it would be foolish to try, as those in whose tradition we follow, to wring a living from this mean land. Though it is not strictly accurate to say that the land is trapped out and logged out, practically speaking it is true when you own a mere acre of it and when the government has finally begun to husband what it had before only administered. As for gardening or farming, the land was gutted long ago by the glaciers and is now a wasteland, agriculturally speaking, though a starkly beautiful one. Thus we are not tempted to live here permanently, reality having effectively quashed that dream even before it had a chance to be dreamed. As long as we are transients, as long as we have great distances to travel to reach Nym, and as long as we have only a limited time to tinker, Nym may be safe in—and from—our hands. For now, we can still say of Nym that it is composed of earth, air, fire, and water.

Caterpillar Nap

Sam Cook

I took my time looking for the spot. It had to be just right. This kind of day comes just once each spring in the North Country. I wanted to savor it. You know the kind of day—after a long winter, the first one when you can feel the sun on your back, the first day you smell the grass again, the first day you can really say it's spring. Never mind that spring came officially three weeks ago. This is the day we were looking for.

I hiked uphill, past aspen and an occasional red pine. I wanted to get over the ridge and out of the wind. It wasn't much of a breeze, but enough to make a difference on this kind of day.

Over the knoll I came to a clearing and looked around. There it was, a young jack pine surrounded by dead grass. I tossed a poncho on the ground and sat down. Perfect.

Not three feet away, a broad patch of snow and ice still held its wintry grip on the grass, but under the jack pine, out of the wind, facing the sun, it was summertime.

The poncho warmed fast in the sun. It smelled like the canvas tents we'd camped in as kids. I leaned back against the jack pine and closed my eyes. The sun baked my face, soaked into my wool shirt, warmed my bluejeans. It was a sensation I hadn't enjoyed since one day on a deer stand last November.

Sig Olson was right. "To anyone who has spent a winter in the North . . . the first hint of spring is a major event," he wrote. "You must live in the North to understand it."

The silence was broken by the cawing of two crows in the distance. Somewhere thousands of feet overhead a jet murmured across the sky. I was pondering the power and speed of the jet when a fly landed on my leg. He seemed to be resting. I stared at him for a moment, then reached out slowly to him. He made no attempt to escape. Finally I touched him. Only then did he take flight. He'll need warmer days than this one if he's to avoid becoming another critter's meal.

Sitting in the sun felt wonderful. Lying in it felt even better. I curled up, resting my head on my arm just inches above the grass. The smell of those decaying grasses and moist roots reached my nose, waking a sense that had been on sabbatical for months.

I opened my eyes and stared into the tiny world before me. Several layers of grass below, a tiny bug was walking down the middle of a green shoot. I watched him for a while before I realized I was looking right past an inch-long caterpillar. The caterpillar was easy to miss. This wasn't one of those fuzzy caterpillars with all the feelers. He was about the thickness of pencil lead and the same color as the blade of dead grass he was crawling on. Like the fly, he seemed to be waiting, gathering strength. It was a long time before he moved, and then it was only a short, inch-worm kind of extension.

I remember watching the caterpillar for some time, feeling the sun on my face, my eyes beginning to close.

I don't know how long I slept. I know the caterpillar was gone when I awoke. The sun seemed lower in the sky. The breeze seemed cool on my neck.

I sat up and jammed my hands in my jacket pockets. I thought maybe it just seemed cooler because I'd been sleeping. I waited. The crows were still cawing in the distance. A few gulls flew over, squealing in their distinctive way. No, it wasn't the nap. The spell had passed.

I stuffed my poncho in my knapsack and headed back over the ridge, home. I followed the footprints I'd left in the snow coming in.

It had been a fleeting encounter, this brush with spring. In another month days like this would seem cool by comparison. For now, though, this was enough. We have learned, Up North, not to be greedy with spring.

RIVERS

Mike Link

Canoe country is not all lakes. The Vermilion River in the west and the Granite River in the east are as wild and scenic as any that have been designated as such. The Rainy and the Pigeon rivers form the border and the backbone of the BWCAW. They flow with high energy over waterfalls and through rapids and widen to form lakes and quiet waters. North Shore streams carve through canyons of volcanic rock, and the Frost River dissects the glacial deposits above Cherokee Lake.

A love of rivers often begins in the fast places where foam and spray break the sunshine into a thousand little rainbows. But soon the edges where green and blue herons stalk become just as fascinating as chutes, eddies, and haystacks.

In the calm waters along the Kawishiwi, the currents are slowed by a reduced gradient and a wide course. Here the river becomes more like a lake or marsh, with shores extended by wet meadows of sweetgale, sedge, leatherleaf, and Labrador tea. Here red-winged blackbirds call their *conk-a-ree* from wind-bent reeds, and painted turtles bask in the sun on logs that are partially buried by river sediments.

The wet meadows hide secretive rails, the shyest of our bird families. The Virginia rail, which is a patchwork of stripes, and the black-faced, yellow-billed sora rail imitate the snipe's sky dance with ascending notes rather than with flight. These aquatic chickens move easily within the shallow water areas and prefer to run rather than fly.

Ducks find places to hide between the wild rice stalks, and black ducks and wood ducks conceal their young in hidden runways. Northern pike spawn in the grassy sections.

The canoe moves sluggishly over the soft, organic bottom of very shallow places. Paddles loosen bubbles of gas from the partial decomposition. These areas will fill up with organic matter and become communities known as shrub bars. Speckled alder and willow will dominate these old stream beds.

The shallow pools between the rapids are the nutrient banks of the river, the gardens of organic matter to support the life that is more uniquely stream-oriented. It is the pool that feeds the rapids, just as the pool regenerates the energy of the paddler before the drop.

As the current gains momentum, ribbons of floating leaf bur reed point downstream and net-weaving caddis flies anchor themselves on the underwater stems. The caddis fly is a terrestrial adult, but most of its life is spent as an aquatic juvenile, with each species developing its own special way of finding food in the fast-moving current.

Because the young are light and soft-bodied, they require an anchored existence to prevent them from being dashed on the rocks and yet allow them to eat. The group known as net weavers secretes a substance that attaches them to the vegetation, and then they weave this glue into a mesh of net that resembles an airsock, with the wide end facing upstream so that the current holds it open. The sock narrows like a cornucopia as it curls under into a narrow foot where the larvae can rest and wait for organic matter from the pools to be swept into the seine.

In the faster riffles, the caddis fly's net would be damaged by the turbulence of the current, so the species that reside here glue bits of stone together to form cones in which they can live. As they grow, they merely add another wider ring of stone to the cones and their houses grow as they do. This cover is not only un-palatable to predators, but it is also ballast to keep the larvae from drifting downstream.

Surprisingly, the concentration of organisms most pe-culiar to streams is found in the rapids. These bastions of fury are also the quiet waters of small life. The myriad stones and rocks that alter the calmness of the surface also deflect the bottom wa-ters into hundreds of small eddies that interact to calm the stream's turbulence and almost halt the current within the min-iature canyons and peaks of the landscape below the white water.

Riffle beetles are well adapted to the current. Their lar-vae are oval and streamlined to confront the current head on. Called water pennies, these young have a body that overlaps their legs and head and sags around them, creating a suction that holds them on the rock. The water penny is a thin suction cup with all its organs and legs inside. Its protective covering provides the animal with peace while it harvests the plant life of the rapids.

The rapids are too strong for water hemlock and the other flowering plants of calm water. They would be bent and bro-ken by the current. But there is life here. There are algae, like the red *Lemanea* that grows only in rapids and falls or the green algae *Cladophora*. Some algae have holdfasts that cling to the rocks, and others have an abundance of mucus secretion that encases the cells and seals them to the rocks. The microscopic desmids live in a world that is beyond our experience. It is a world of float-ing and drifting, so free and light that the mere surface of a rock is such a force of friction that the desmid sticks to it rather than flowing with the current. The rocks are often covered with thou-sands of desmids or mucus-secreting algaes, and canoists slip and bang their shins when they try to walk their canoes through these areas.

The froth that is white water is a whipped mass of gas and liquid, water and air. This oxygen-rich environment is less dense and less stable than normal water. It is here that the ka-yaker can paddle a submerged boat, because it is less buoyant in the lighter foam. It is here that the water of the river takes on its oxygen load and purifies itself.

The animals of the rapids depend on this oxygen rich-ness for their survival. If the stone fly's search for food takes it too far from the main oxygenated water, it will pause and do "push-ups" to increase the flow of water through its gills.

Pulsating black masses line the walls of the northern waterfalls—millions of blackfly larvae attached to the wet, slippery rock face. Attachment is frequently the key to survival in this aquatic world. The blackfly larvae reside in the fastest waters be-cause they need the oxygen of falls and rapids. When they become alarmed, they drift downstream and then, when the coast is clear, return via a thread that has been spun from the rock. Unlike the midge and stone fly, the blackfly emerges in fast, full streams. When it has reached maturity, the back of the larvae casing splits and an air bubble emerges with the adult. It floats to the surface, dissipates, and the adult flies away.

Fly fishermen know that a good lure placed in the cur-rent just at the edge of the rapids and allowed to float into the pool beneath is a good way to catch a hungry trout. Fish line up be-neath the rocks, with their heads toward the current. The water breaks gently, flows around them, and forms eddies behind their tails, allowing them to wait without effort.

The trout are waiting there because, inevitably, some-thing will happen to the otherwise secure rapid's life, and the or-ganisms that live there will be cast into the current and sent downstream. Those who study the river life call this drift material. Drift material is life caught in a maelstrom of current and can be generated by a curious naturalist overturning a rock, a surge of water from a storm or dam opening, or a rolled rock caught on the keel of a canoe. Underwater, the surge of energy is sudden and strong, and life is ripped from its hold in a chaotic moment of disorientation. The eddies are overwhelmed, the miniature can-yons feel pulses of current, like flashfloods, and those organisms caught in the fury are swept into the pool below and the net of fish mouths that wait for such a moment.

The river seems to lose some of its width as willows and alders lean out to reach the sunlight between the shores. They secure the shore from the cutting power of the current and in the summer support the warblers and flycatchers that feed on the emerging insects. It is in this zone that the kingfisher ties the land and the water together, nesting in the ground, hunting from the branches, and taking its food from the water with a quick flight and the stab of its beak. Kingfishers move downstream from the paddler, rattling and diving as though sewing land and water together with stitches that reach from branch to water and back to branch again.

Seasons pass and so do years. Bloodroot and marsh marigold lead to boneset and joe-pye weed, running waters give way to oceans, adult insects emerge and return to lay eggs, and life derives energy from death. There is always change, but there is also continuity, and the river reflects the health of the planet and the joys of discovery.

Oriental philosophy draws many parallels between the river and life, from its small beginnings to its slow and ponderous end, with rocky spots and smooth stretches in between. The river is a symbolic organism of strength and moods (floods), changes and adjustments (oxbows), growing with each new encounter (confluence) to become the sum of all its experiences before pouring them into the collective knowledge of the planet (the oceans).

For those of us seeking the essence of the river, there are many ways to experience its breadth and length. To those who would know the river best, the experience must be both physical and intellectual, for the sum of the two is wisdom. The river is the naturalist's highway.

ROCKS AND LICHENS

Mike Link

The rocks of canoe country are significant to all paddlers. They set the tone for the paddle. At times, the rocks tower overhead, dwarfing the paddler from their loftly heights, while on another lake, the rocks seem to be gone, until the keel slides onto a giant lurking beneath the water and reaching from the depths. Rocks attract us for swimming and sunbathing, for lunch stops, and as havens from biting insects. The rocks are more than remnants of Precambrian time or glacial sculpturing. They represent our human idea of permanence and strength. Their size, shape, and structure are fascinating, and the colors of the pink granites, greenstones, and dark gabbros contrast with sky, lake, and forest to paint pictures on the lake's reflective surface.

In the Saganaga and Seagull lakes area, a pink granitic rock dominates the islands and shoreline. It is a rock that formed beneath the surface and contains quartz aggregates, marble-size accumulations of quartz that are circular in shape. This is special rock, found nowhere else. It is part of the story of the earth, of molten magma, of old mountainous uplift, melting, and freezing. Nearby are other spectacular rocks that inspire geologists and carry their imaginations back billions of years.

Along the Gunflint Trail, an iron formation, banded and oxidized, sits as a road cut. It is ancient rock, a concentration of chert and iron that eroded from ancient landforms and collected in an early sea. Nearby is a trail to the magnet rocks, magnatite. The trail winds through the boreal landscape and ends at a large rock sculpted from the landscape by the last glacier.

There is a sense of monument about these rocks, something dramatic and mystic. They affect the compass and render it unusable in this area. A look at the lines on the topographic maps attest to the geographical confusion. The township and range lines waver in and out, creating sections of unequal distribution. These were set by early surveyors who didn't have airplanes or nonmagnetic survey devices.

West of Saganaga in the Knife Lake area, a conglomerate rock dominates the cliffs. Rocks embedded in rocks are perched in horizontal layers. These were cobblestone beaches when the land of Saganaga was higher and eroding into ancient seas. The action is stopped, held in a timeless form that weaves a story for those who pause and ponder.

Duluth gabbro is the dark rock of canoe country, an intrusive rock filled with large crystals like a granite, but with crystals that are all dark minerals. These are the rocks that make the most dramatic cliffs. They represent a time when our continent was just forming, a time when it nearly split in two and a large rift extended from the BWCAW to Wichita, Kansas. Lava from a chamber beneath the present Lake Superior spilled out of the crack and made up the surface lavas near the shore. Then as these lavas cooled and plugged the rift, another surge of magma squirted beneath the surface to form a series of large gabbro sills.

As I canoe beneath this scenery, I absorb more than beauty, I also learn perspective.

The early canoeists must have seen more than rock, too, or they would not have been inspired to leave paintings on the sides of the cliffs. It was no idle thought that put the graffiti on the rock. And it wasn't done with a spray can, but with a mixture of red ochre and fish oil that was mixed together to create a permanent record. An ancient voice was placed on the rock.

The rocks hold other artwork as well, artwork that varies over the centuries with growth and competition. It is the design of the lichens, hardy pioneer plants that tenuously survive on seemingly impenetrable rock. Lichen comes in many forms. There is a crustose type that is embedded in the rock, its phenol acid etching the surface in a way that allows the plant to grow intertwined with the crystalline structure. These are sometimes called coin lichens for the concentric growth pattern. The common ones in the BWCAW are gray and greenish.

The most exotic group is the fruticose lichens, a group with fruiting bodies that mimic the fungus within the lichen. Lichens are a partnership that is sometimes called a symbiotic relationship. Two plants are united as one. There is a simple algae that can reproduce by splitting and can provide food by photosynthesis. Because the algae would dry up and die or slide off the face of the rock, a fungus part of the lichen has little threads that can grip the rock. The rock then deteriorates under the solution of phenol acid the fungus releases. The fungus can hold on, and it can form a hard exterior skin with a spongelike center that can hold water. It cannot make food, however, and the rock has no food to give. Therefore, the lichen must depend on its algae part for sustenance.

The shape of these lichens depends on the form of the mushroom that the fungus would produce if it were independent. We find light-gray reindeer lichens, light-green forked lichens, funnel-shaped pixie-cup lichens, organ-pipe, ladder, awl, deformed, and numerous other lichen variations. The most colorful is the British soldier, which resembles match sticks with red heads (coats). All of these were the food of the caribou. They grow under jack pines and on open rock slabs. They are pioneer plants and old forest plants. Variations of this type hang from trees as old-man's-beard, a lichen that the parula warbler weaves into a nest.

The third type of lichen is foliose, a leafy plant form. One group is the umbilicates, with single large, leathery leaves that are attached to the rock by a group of fungal threads that are clustered in one umbilical mass. These are the tripes. Most large, vertical rock faces have them. They come in brown and gray, with the gray ones looking like elephant ears when they are wet and pliable.

The other foliose group forms shields, round rosettes of green, gray, yellow, and orange. They dominate the rocks of the canoe country, especially the bright, reddish orange rock lichen. Like the rock pictographs, this reddish orange coloration reaches along the cliff faces toward the water, where its reflection stretches like a train across the mirrorlike lake.

The colors of the lichen are bold, and they are set off by the dark forest above and the rocks that support the growth. No plant says canoe country more to me than this one, no plant represents the struggle against the demands of the north country more than the lichen, and none brings the land more cheer.

VOICE OF THE ROCK

Milt Stenlund

A hundred centuries ago and more, the land reached stark and sterile to the horizon and beyond. Rounded hills of solid glacier-scoured granite appeared in all directions with broken bits and boulders at their bases. Scattered about and covering much of the bedrock were piles and drifts of assorted gravels and sands often mixed with darker soils. Only the softening greens of pioneering willows, stunted aspens, and dwarfed birches struggling for foot-hold in the more fertile soils broke the neutral colors of the raw landscape. Even the icy cold waters of the nearby lake were gray-blue, colored by fine powders ground from the rocks for eons by billions of tons of ice.

On this warm summer day, however, heat waves hung over the calm waters and minute mirages hovered and disap-peared over the mirror surface. The only sound was that of mov-ing water—hundreds of tiny streams trickling busily from the ledges and gravel moraines to finally splash into the lake below.

On the south shore of the lake and perched precariously on a long sloping granite ledge was a massive block of rotting glacial ice. Protected for centuries by a heavy mantle of gravel that insulated it from the heat of summer, the ice block clung to the north-facing slope long after the edge of the giant ice sheet had retreated north.

Wild sedges grew in the gravel and sand banks, and willows had occupied the shoreline to the south for decades. Now the probing summer sun had caused deep gouges in the ice surface. Crystals and slush appeared where once the ice was clear blue, and small heat-absorbing rocks were sinking further into the old glacial remnant.

Suddenly there was a thundering report that echoed across to the cliffs on the far side of the lake and carried rebounding down the lake. A large crevasse appeared near the top edge of the ice block. Instantly a jagged streak ran the full length down to the water's edge. The split widened, slush and torrents of water cascaded over the sides, carrying broken rocks and rounded boulders. The two halves parted, slid down the steep grade, and with a grinding roar splashed into the lake, sending a series of tidal waves across the waters to crash against the far shores. For several minutes water and gravel continued to tumble down the slope and large boulders, released from their icy prison, rolled into the lake. Then all was still.

The wet pink surface of a long granite rock ledge lay fully exposed to the sun. Scratched, carved, rounded, and gouged by slowly moving glacial ice for thousands of years, the bare rock was finally released from its burden and opened to light. Formed a billion or more years earlier by massive volcanic convulsions that forced the molten materials to the surface, the ledge cooled slowly and was fractured, bent, melted, and cooled again by succeeding earth-making processes. Now the early history of fire, heat, and pressure was preserved forever in the convoluted bands of dark and light rock and the white quartz dikes formed when later molten liquids were forced into cracks of older rock. Earlier glaciers and centuries of erosion had worn the softer rocks down so now the white dikes were raised slightly above the rest of the surface.

For a century there was little change in the raw rock surface. A hundred winter snows covered the ledge and a hundred summer suns followed to warm the rock and assure that the recent ice age was over. Winds blew over the ledge and thunder and lightning rolled erratically across the summer skies, but the most persistent voice of the rock was the gentle whisper of waves as they entered a narrow crevasse that began several feet below the water surface and ended abruptly a few feet above water line, where it met a hard quartz dike. The voice varied with the weather—still and resting while locked in frozen winter and raucous during the violent storms of summer, which sent white-capped waves crashing along the shores. Then small geysers erupted and spumes of wind-whipped spray blew out of the crack and over the ledge. The gentle voice was overwhelmed by the sounds of rushing waters, lightning, and roaring winds.

But mostly during the warm summer days when small ripples raced across the surface of the lake, the voice sent out its distinctive sounds, gurgling with the ripples and slapping rhythmically with the choppy waves. At night when all was still, the voice was reduced to a whisper as water welled up and down slowly in the dark cavern of the fissure.

One day in late spring, a single blade of wild sedge appeared in a small crack filled with fine sands washed in from above. Wind-blown spores next settled in a shallow area nearby and small rock ferns appeared to fasten their tender tentacles to the rock. Later more spores settled on the roughened surface and acid-secreting lichens took hold and spread slowly.

The ledge surface again changed little for a period of centuries. More of the ledge was exposed during periods of long drought and the lake water reached higher when rain was plentiful. Sedges and grass now occupied several of the sand-filled niches and rock ferns colored much of the shaded north-facing side. The lichens had spread slowly and the raw appearance of the rock had darkened to gray and black. Only the quartz dikes remained white. Alternate freezing and thawing had broken off several small pieces, which lodged further down or dropped into the lake. The main crevasse from which emanated the voice of the rock remained unchanged.

Major changes took place in the hills above, however. The small sandbars had been washed into the lake or were now covered with heavy growths of willow. Hardy pioneering birch trees occupied much of the newly formed soil in the valleys and black spruce was growing in the poorly drained lowlands. A thousand years passed and then another. During a long warm period of several centuries, oaks and hickories appeared on the hills. The lake level dropped so low that the main crevasse in the rock was almost completely exposed. The subdued voice was again reduced to a whisper. Then for a period of years the cool weather returned and rains fell. Spruce and fir reappeared on the slopes and the oaks disappeared, leaving only their spring-dispersed pollen to sink into the depths of the lakes.

The ledge changed little. Lichens now covered most of the surface with grays, greens, and black. Small frost cracks drew geometric patterns, which were filled with dark green hair cap mosses, and an edge of crusty caribou moss reached down from the woods at the upper edge. During the summer sandpipers paraded along the water's edge and white-footed deer mice ventured at night from the security of a small aspen grove. In the winter bands of woodland caribou traveled the lake ice and early one fall a bull moose swam across the lake, scrambled up the ledge splashing sheets of water and disappeared into a sphagnum moss bog in answer to a bawling cow. A white-throated sparrow sent its plaintive notes from the spire of a balsam fir and a loon answered far down the lake. The voice of the rock, however, remained unchanged.

Eight thousand years after the ice block had dropped into the lake a new form of life appeared. On a hazy day in late fall when golden aspen leaves covered the ground in the valleys and orange and red lichens and mosses colored the otherwise drab rock, two human figures emerged from the woods above and walked down onto the ledge. The man carried two hindquarters of freshly killed caribou calf and the woman the liver and heart wrapped in the skin along with some carefully selected bones. While the man rested near the water, the woman unrolled the green hide and began to carefully scrape off the fat with a crude stone fleshing knife. When finished, she rolled the hide into a bundle and while she was wiping the stone on some moss nearby, it suddenly slipped from her grasp and dropped into the crevasse, disappearing in the black depths. She peered intently into the opening for a moment, then stood up, shouldering the bundle.

The two figures talked briefly and then moved slowly down the shoreline, heading south to their winter camp. Aboriginals had visited the rock.

The years moved by, the forests above changing slowly with differences in rainfall and temperature and rapidly when fires started by lightning roared through in dry years. Several centuries after the first human being had rested on the rock, people again appeared. On a midsummer day, a birchbark canoe with two paddlers appeared from the south. It worked its way along the shoreline and then after a brief verbal exchange and some pointing at the hills, two Indians nosed the canoe carefully along the ledge and stepped out. They were searching for ripe blueberries. The woman disappeared in the woods while the man worked at placing a small birchbark patch on the canoe where a sharp underwater snag had ripped the bottom. He then sat back to wait for word from above. While stretched out on the ledge, staring across the lake, he became aware of the sounds from the crevasse—the quiet whispers and voices repeating the story of the rock. The Indian bent over the opening and then, putting his ear down, he listened intently until the woman appeared to report that berries were scarce, not worth the picking. The man pointed at the crevasse and said, "Gay tay ah nee she nah beg—people from way back." He then reached into a small buckskin bag hanging at his waist and carefully sprinkled several pinches of kinnikinnic tobacco into the darkened opening.

Decades later an entirely new sound was wafted across the blue waters to echo against the gray cliffs above the ledge. The rhythmic sound of a chant and flashing wet paddle blades accompanied a canoe of French voyageurs as they sped down the long reach of the lake from the east. At a command, the speeding craft suddenly turned and headed for the ledge. It was time for a smoke. As the prow touched shore the occupants scrambled out, filling the air with laughter and jibes at the helmsman. A pack was opened and passed about while a small fire quickly heated a brass kettle for tea. One of the men, more reserved than the others, walked across the ledge to sit down with his pipe near the crevasse. While he puffed slowly he looked across the water and into the far horizon over the pine-covered hills to the west. At the same time he became aware of the subtle whispers that rose from the recesses of the narrow opening. When the order came to reload, the voyageur tapped the cold pipe on his hand and then leaned over to put his ear to the hole. He listened intently for several moments and then as he rose, the white clay pipe slipped from his fingers and disappeared into the dark water. He frowned momentarily, then smiled, raised his hand in a half greeting, and turned to help the crew.

That winter a pack of timber wolves streamed out onto the ice from the far shore in chase of a small band of caribou that had been walking slowly down the lake ice. They quickly headed for the near shore and the safety of the deep snows in the woods. The wolves gained quickly on the packed snow of the lake, but the caribou reached the shore and headed for the cover of the trees. The last animal faltered, however, as it scrambled up the snow-covered ledge and those last seconds proved its undoing. After feeding, the wolves climbed halfway up the slope and bedded down in the shelter of a small stand of pines. The next morning they returned to feed again and then headed single file down the lake, resuming their mile-consuming trot. Several ravens and an eagle fought over the carcass for days and then a fisher dragged a leg bone up the ledge and chewed on it briefly. During a heavy thaw the following spring most of the snow melted from the rock in a day and the leg bone dropped into the still-frozen crevasse.

The face of the ledge had changed little in the past several centuries. The slow processes of erosion had roughened the surface and rounded the original sharp edges of the crevasse. Lichens, liverworts, and mosses covered much of the upper areas, and crusty caribou moss now extended down from the forest edge to form a crisp covering during the summer months. On hot August days the pungent yet pleasing smell of sweet fern was wafted across the ledge by gentle winds. Only the ice- and wave-buffeted surface near the water's edge remained free of plant growth, but it had darkened in color so that the original pink of the granite was no longer visible. The sloping ledge had now been exposed to weathering processes for more than ten thousand years. The forests above the lake had changed again so that now giant red and white pines covered much of the terrain. Two centuries earlier, following a long period of drought, a massive fire fed by high winds had swept across thousands of square miles of the land. The heat was so intense that the fire created its own tornado draft, sending burning bark and branches hundreds of yards ahead of the roaring inferno. The ground duff and humus that had been accumulated for centuries was reduced to fine ashes and blown away, leaving gravels and sands exposed again. Here the scattered still-standing red and white pines found a receptive mineral soil for the millions of seeds that were dropped. The history-making Lake States Pinery had taken root, eventually to cover hundreds of square miles with the finest of lumber-producing trees.

Only the aging voice in the rock remained unchanged. The dark crevasses harbored no green plants and the annual scrubbing action of winter ice kept the rock surfaces clean. The voice had whispered, cajoled, and scolded now for a hundred centuries. On the bottom in the icy depths of chilling water lay a crude stone shaped for fleshing hides, a ghostly white clay pipe, and the yellowing leg bone of a caribou. The voice guarded its ancient secrets well.

On a cold day in late fall a man suddenly emerged from the pines at the very top of the hill high above the gray, choppy waters. He looked frowningly over the white-capped lake and into the racing snow clouds to the north. After lighting a short blackened pipe, he adjusted a small haversack at his side and then joined two companions who were cutting witness corner blazes on nearby pines that surrounded a small carefully built cairn of rocks. After a careful compass reading and a brief discussion, the three left, heading for the west end of the lake. The moss-covered ledge, the dark crevasse in the rock, and the wind-whipped lake were suddenly and efficiently moved from unrecorded ancient history to the present in the worn leather notebook of the government land office surveyor. The cairn of rocks marked the blazed township line which came up from the south. Carefully written notes described the lake, the islands, the rocks, and the timber. The magic of the written word would now make the land known to persons throughout the nation, hungry for land, timber, and ores.

The pace of discovery quickened. Short years later witnessed the invasion of hobnailed boots, double-bit ax, and wool-clad lumberjacks. They arrived from the south end of the lake on a small steam engine–propelled barge and immediately set to cutting short haul roads up the valleys into the pines above. All winter long the crews of lumberjacks worked in the hills, arriving before sunrise and leaving as the sun left the sky in the evening. Horses skidded the timber down the iced trails and left it piled in windrows on the ice in front of the ledge until the entire bay was covered with resinous logs. In the spring it was time to move the winter harvest to the end of the lake. A wooden dam had been constructed the summer before in the outlet of the river and, with the gate closed, the lake level gradually rose until trees along the shore were flooded and almost half of the crevasse in the rock was filled with water. The voice was subdued but not overcome. On calm days the words were almost inaudible, but when the white caps appeared and the wind was from the north, the voice again related the history of the rock.

Halfway up the long ledge lay a pile of a dozen logs, each more than two feet in diameter and twelve feet long. On the day the raft of logs in the lake below was to move, two men came up to work on the key log in the pile. It suddenly broke loose and rolled sideways while the others tumbled down the ledge to splash into the lake. The jacks shouted and jumped nimbly aside but an errant log rolled over the double-bit cutting ax, snapped the handle as a twig, and sent the steel head tumbling down the ledge until it disappeared with a splash in the crevasse along the shore.

By fall all the pines worth cutting were gone from the hills and valleys. Only the crooked, double-top, or hard-to-reach trees were left standing as lonely sentinels. Late the next summer a fire (started from sparks from an engine some distance away) swept quickly over the cutover hills, burning the dried brown branches and pitch-laden treetops left as slashings on the ground. Once again the hills were reduced to a desolate waste of stumps, rocks, and ashes. Except for a few raw scratches from "corked" lumberjack boots, the ledge and the crevasse remained unchanged.

For the hundredth time the inexorable processes of natural healing began on the age-old hills. Royal blue mantles of blueberries covered much of the ledge and burned areas. Pin and choke cherries appeared almost miraculously and sprouting birch, aspen, and willow again covered the glacial moraines with quiet pastel greens. Snowshoe hares increased by irruptive numbers in the brushy areas and white-tailed deer roamed the valley by the dozens where once the moose and caribou held reign. Beaver reappeared in the lake and built their lodges along the shoreline, venturing only short distances into the valleys for high-preference aspen for food. The newly grown woods were alive with the songs of thrushes and warblers and the constant chatterings of red squirrels and chipmunks. Billions of aspen leaves trembled in every breeze and bloodred maples added colors in the fall. It was a time of burgeoning life on the land—far different from the quiet of the pines and their soft winds.

The woods remained productive for four decades, then subtle changes once again began to have their effect. No fires had occurred and the only evidence of previous burns were the fire-scarred pine stumps and occasional blackened and rotting pine logs and standing snags. The aspen and birch were now mature and their heavy crowns of leaves gave almost complete shade to the ground. One by one the sun-loving and short-lived cherry trees died and fell to the ground. Annual plants that once provided bounty to hares, deer, and grouse were replaced by the evergreens of bunchberry and wintergreen. The woods assumed a parklike appearance and the once abundant deer browse species declined, to be replaced by hazel and upland alder. Fewer deer now appeared along the woods' edges and snowshoe hares found little to eat close to the ground. The once lush postfire brush was gone or grown out of reach. Horned and snowy owls found good hunting on remaining hares as the cover disappeared. Few new broad-leafed trees were replacing those dying above. Instead, small balsam fir and spruce seedlings, able to survive in the shade, were sprouting by the millions. The spruce-fir forest, well suited to this northern climate and soil, was again taking foothold through natural succession, just as it had times earlier when conditions were suitable.

On a bright day early one summer, a small wooden boat coasted along the shoreline to the ledge and into the small sand-bottomed bay nearby. A family of four scrambled through the thin line of tag alder along the shore and walked into the small parklike stand of aspen and birch just beyond. To the east sheer cliffs rose up to the sky and to the south lay a small cedar grove with signs of winter use by deer. They talked excitedly for some time, laying out an imaginary cabin, and then the family walked out to the ledge to sit down. No words were spoken as they drank in the beauty of the lake and its far shores, then suddenly they were drawn to the crevasse and its sounds. Whispers and chortles tumbled out endlessly as the waves entered at water level and broke on the unseen rocks in the dark cavern. They leaned over to listen closely and the man said quietly, "It's the voice of the rock. Listen closely and you will hear the story of the lake and the rock and the ancient land. You will learn of the storms and gentle winds that have blown a thousand years, of the changing forests and the fires that destroyed them. Of the ancient red man who hunted these hills and the timber wolf that stalked the caribou. It tells of volcanoes that spewed molten rock and of tiny icicles that hung from the ledge on a spring day. It tells of the centuries-old struggle these lichens have made to gain foothold on the granite surface and of the lush growth of bountiful foods provided by the second growth forest after a fire. It tells of the changes of a thousand centuries and yet this rock, this ledge, and the voice have remained the same. It is the story of this harsh yet beautiful land, our land, that lies west and north of the great lake called Superior."

Navajo religion is not concerned with life in another world. It is based on man's efforts to achieve a balance between himself and nature, to obtain perfect harmony with the forces around him.

Navajo Tribal Museum
Window Rock, New Mexico

QUETICO-SUPERIOR

William O. Douglas

My island is a special place. It is mostly rock and less than an acre in size. It has only a few trees. Yet it holds a warm place in my heart. It represents to me the great northland that stretches from Lake Superior to Hudson Bay, from Fort William to Winnipeg, and on to Athabaska and the great Mackenzie that flows to the Arctic.

My tiny island is the shoulder of a granite mass polished smooth by glaciers that slid slowly for centuries down from the north. The thick ice ground well, thousands of tons beveling the edges of this island as though it were a jewel. It is a grayish-green stone set in deep blue. A canoe touches it gently, for there are no rough edges. The polished rock disappears quickly into the depths to mark a pool where smallmouth bass and walleyed pike cruise.

This rock is typical of the Canadian Shield, a vast area of gneiss and granite. It is shaped like a U, with Hudson Bay in the center. It touches the Arctic Ocean and beyond; it also reaches at points into the United States. The glaciers pushed much of the choice soils south, leaving barren, sterile rock. This rock was first invaded by caribou moss, which is a lichen that decorates this little island.

The word *lichen* comes from the Greek, meaning "leprous" or "scaly." It is, as Donald C. Peattie once wrote, life in a new biologic dimension—two organisms united as one and indissoluble. Half is an alga, the other half a fungus. The alga does the photosynthesis—that is, supplies the food—while the fungus supplies the water. A lichen lacks stems, leaves, roots, flowers, and seeds. It consists of a tightly woven web of fungal threads, or hyphae, through which colored algae are scattered. They reproduce by fragmentation. A minute spherical body of hypha may be blown away. A few green cells of its alga partner travel with it on the winds. Cities with industrial development are devoid of lichens, for ozonated hydrocarbons kill them. They flourish only where the air is pure. They are found in jungles and in deserts; they constitute most of the life in the Antarctic; I found them on all the continents and as high as 18,000 feet in the Himalayas.

They grow thick and almost lush in these northern woods, since atmospheric moisture is in great supply. Even when the snow comes, they continue to grow. They are probably the slowest of all the perennials, their best growth taking place when the temperature is almost twenty degrees. Some lichens prefer conifer bark, some prefer hardwoods. Oak-hickory, oak-pine, beech-maple forests have different lichens. Lichens also have their preferences. When it comes to rocks, some prefer granite, some basalt, and so on. In this north country they appear in massive quantities to form soft gray-green carpets on the ledges.

They had transformed my island. The spores had taken hold in tiny fissures or abrasions, sending out roots that produced acids that either dissolved the rock and widened the cracks or broke off chips of rock by expanding or contracting. This lichen, which contains carbohydrates and which is a mainstay of the woodland caribou, gradually spread and conspired with time and water and frost to enlarge its domain and possess the rock. Soon humus—now several inches deep—began to form; and gradually it filled the larger cracks. The floral invasion now assumed more formidable proportions. Grass and shrubs grew from the humus. They soon occupied half of the island, blueberries and wild strawberries being predominant. Then came the trees. Tiny seedlings of pine found footing. Some died; others lived. Several grew for centuries, prying the rock apart, sending their hungry roots deep for nourishment. When the voyageurs who came this way in the seventeenth century pushed their birchbark canoes past this island, the dead-white pine that now stands gaunt and naked was flourishing. Jack pine had also taken hold and then perished in midlife. New trees had followed—ash, jack pine, and white pine—putting my island in partial shade. A few small willows about two feet high decorate the fringe.

Thus does my island represent in miniature the emergence of the northland from a place of sterile granite and greenstone to rolling land decorated with stately trees and blessed with choice shrubs. My island, like many other points and headlands, represents a place of solitude bounded by blue waters where beavers splash, loons call, and ducks feed on wild rice.

This island lies midchannel a hundred yards or so below the Lower Falls of Basswood River. Four falls fill the place with a roar. Torrents of water boil with rage, forming a mist. No canoe could survive them. Granite ledges run across the middle section, sending the waters into angry sluiceways. I saw the island first in moonlight, when Sig Olson and I camped upstream and paddled down to see the falls. That was at the start of our hundred-mile canoe trip. When we portaged those falls the next day, the full floodlights of the sun were upon the island. A thick mist was rising from the waters. I then knew that the island gave choice seats for the great wonders of the northland. That is why when we finished our great circle and returned to Basswood River, Sig and I decided to make this our last camp.

We got wood from the dead snags on the little island, and built our fire in a large indentation in the sloping granite rock. We brought birch poles from the mainland and pitched our tent. Two casts with a spoon into the water below the falls brought a walleyed pike to shore. Sig served its steaks with sweet, crumbly corn bread. Dishes were done, beds were made, the canoe was high and fast for the night—all within two hours of the time we first landed here.

Then a full moon rose, and the roar of the falls, which I seemed not to notice during our busy two hours, suddenly filled the place. A chill settled over the waters. A thin mist started to rise at the foot of the falls. Moonbeams danced on waters at the head of the pool. A loon called somewhere in the darkness behind us. I put wood on the fire, buttoned my parka, and relived the journey.

Not far below this island we had stopped in Crooked Lake to see the historic Arrow Rock. The pine and spruce were somber. But this was late September, and color had come to the north. Touches of red from oaks and maples and splashes of yellow from aspen, birch, and poplar lifted the heart. Every promontory, every bay brought a fresh, vivid view. Some shores were solid gold where groves of birch grew thick; some hills showed yellow or red only on their crowns. Every view was a gay one.

Arrow Rock is a cliff a few hundred yards long and perhaps a hundred feet high that drops straight to the water on the upper end of Crooked Lake. It gets its name from an event that is probably not legendary. This lake country was the land of the Chippewas and often invaded by the Sioux. Once the Chippewas evaded their enemies, who hunted for them in vain. The Sioux on their retreat stopped at Arrow Rock and shot their arrows into a long diagonal crack across the face of this granite cliff. Thus they showed their derision. The story gains credence from the early chronicles of travelers who report finding arrows there. But they had disappeared by 1850. Today Arrow Rock is famous for the pictographs painted there centuries ago. The paintings evidently mark the site of an old camp where the Indians could paint in leisure. Red paint was used here, as it was in most of the pictographs we find in the Pacific Northwest. The red paint was made by grinding iron ore and mixing the powder with fish oil. A thin solution of resin from pine or spruce trees was added to give the paint a varnishlike quality and make it practically impervious to weather.

The paint on Arrow Rock is bright red to this day and gaily outlines moose, caribou, and pelican. The pictographs also include round red disks apparently representing the moon. We brought the canoe alongside the rock to photograph the paintings. Then, beaching the canoe on a polished granite reef, we climbed to the top. Jack pine was flourishing in the granite. Here and there were mountain maple standing no higher than a shrub, their leaves bright red, their smooth bark reddish-brown. Blueberries and bearberries had found footing in cracks. Gray-green caribou moss lay in large patches. Other mosses that the caribou like were scattered along the top, making an exquisite carpet with a soft, thick nap and somber color. On the side of the cliffs was a black lichen shaped somewhat like a saucer and attached to the rock at its center. The voyageurs called it "tripe of the rock"; and it has some food value, cooking down into a mucilaginous dish.

Sig and I sat in silence, drinking in the solitude and grandeur of the place. The deep blue waters of Crooked Lake were calm and placid at our feet. The dark-green forest touched by bright-red and brilliant yellows reached as far as the eye could see. This rolling country has few ridges and no peaks. Now and then a sheer cliff showed; a distant touch of blue marked another of the thousands of lakes that distinguish this wilderness country. The depressions gouged out by glaciers filled with sweet water to make the most extensive lake system on this continent. The waters stay fresh, draining a vast area. Sig broke the silence to explain that the drainage in this part of the Canadian Shield was unique—a low dome with the surface sloping from its center to the edges. The waters from this rounded glacial plateau, mostly under two thousand feet in elevation, run to three points of the compass— north, east, and south.

In the early years this was almost exclusively pine and spruce country. Its swampy lands, rimmed by cotton grass and filled with bulrushes and lilies, were ideal for moose. The lichens that the woodland caribou eat go with spruce and pine. Thus this vast lake region was part of the range of that arctic animal. When the lumbermen came and started cutting, the vegetation changed. Aspen, oak, birch, and maple replaced the pine and spruce; and the caribou moved north. That trend was accelerated by the fires that haunted this area once humans came. On my way north to Ely, Minnesota, where our trip started, I had passed through the prairie town of Hinckley. In the early 1890s it had been a lumber town. The prairie was then densely covered with white and red pine. A long drought made the fields and forests like tinder; and in the late summer of 1894 a fire broke out. Hinckley had protected itself with a hundred-yard clearing that ringed the town. But the fire came racing along the crowns of the pine trees at sixty miles an hour and crossed the clearing in one jump. It killed 275 people in Hinckley alone.

This second growth of aspen, poplar, birch, and maple that comes in after a fire makes wonderful browse for deer. And Sig explained how, as the caribou retreated, on the disappearance of the pine and spruce, the deer moved up from the south. The cycle is changing, now that the area has been stabilized as a wilderness. About two million acres are guarded against the saw and the ax and patroled for fires. Pine, spruce, and balsam are reclaiming the land; and some caribou are drifting south. But the splotches of red and yellow that adorn the somber northland in late September will be there to brighten the lives of those yet unborn.

We turned east on Crooked Lake and passed through Moose Bay to Robinson Lake, making a pleasant portage. Beavers had built a dam across the outlet of Robinson Lake, raising the water level a foot or so and creating a pond that was important in the ecology of this country. There ducks would find refuge; deer and moose, forage. We sat on the shore, watching the beavers at work. We were not alone, for a great blue heron standing on one leg was also a spectator. The beavers had built the dam to raise the water to flood stands of aspen and birch that otherwise they could reach only by land. Now they were busy building canals into the newly acquired stands and bringing back choice branches for their winter's food supply. Soon their house at the dam would be eight feet or more high. They would work in comparative safety, as few predators would reach them if they stayed in the water. Their chief predator—human beings—was now effectively restrained. For only Indians were allowed to trap on the Canadian side of this wilderness region; and few of them do so.

"That's the ideal life, being a beaver," I said to Sig as we put the aluminum canoe into Robinson Lake.

"Not so," said Sig, who is one of our foremost experts on the northland. He went on to explain that great tensions develop in animals as well as in people. Studies made at the Wilderness

Research Center (which Frank B. Hubacheck of Chicago founded on Basswood Lake and which is run as a cooperative project by several universities) show that, once the population of mice and shrew mounts on the mainland, the animals move to the islands, crossing in the winter on the ice. I had learned of this phenomenon when I visited the Brooks Range in Alaska. I had heard from the lips of Olaus Murie and Sally Carrighar the story of the migration of the lemmings. Overpopulation of this member of the mouse family creates a shortage of food that produces panic in the animals. Hundreds and thousands of them strike out overland for a new home. It seems to be almost a suicide pact, for they go on and on, crossing rivers and being washed away, reaching a lake and drowning in it, or actually jumping off cliffs into the sea, once the ocean is reached.

Other animals have like tensions. Rats, mice, muskrats, rabbits, and deer also know the deadly effects of stress. And Sig told me, as we paddled slowly across Robinson, of the study Lowell Sumner made of beavers. When the beaver community is young and newly established, there is food for everyone, homes for all the families, work opportunities unlimited. Young beavers raised in that environment have wide horizons; there are stands of willow, aspen, and birch to be flooded, dams to be erected. The aggressive predators are at bay, for the dams and the canals built into the newly flooded areas of willow, aspen, poplar, and birch give good protection. As Lowell Sumner states, "For beavers this is history's Golden Age of self-expression, of freedom from want and fear." As years pass and the population increases, the food supply per beaver diminishes. Young beavers leave home in search of new frontiers, new food supplies, new opportunities. Those that remain have to work harder for their daily food, travel longer distances, get fewer rewards. Diminishing supplies of willow, aspen, and birch mean that existing dams cannot be repaired properly. The result is a lowering of the water level and a further contraction of the supply line. Malnutrition increases; as it is prolonged, fertility declines; infant mortality reaches an all-time high. But the greatest killer of all—more potent than malnutrition, disease, and predators—is stress. When that reaches its peak, the beaver population drops at once, vast numbers dying quickly.

Lowell Sumner has shown from postmortems of beaver the damage that stress has done—inflammation and ulceration of the digestive tract and permanent metabolic derangements.

"The beaver is like man," Sig added. "Stress is the greatest killer of all."

Those thoughts lingered long. Researches on the beaver placed in new context for me the values of the wilderness to us; they reemphasized the price we pay for civilization.

Every wilderness trip has its shining moments that persist in memory when all else is only dimly recalled. One such was the few hours Sig and I spent on Sarah Lake. Up to that morning the sky had been overcast, the air heavy. The north woods seemed gloomy except as the red and yellow leaves of the deciduous trees lighted up the landscape. But this morning as we finished our portage out of Caribou (Tuck) and across to Sarah, a west wind picked up and the sky brightened. Dark clouds fled before the freshening gusts. Now the waves sparkled in the September sun and fleecy clouds raced across the sky. A new spirit seemed to possess the canoe. Now our paddling was not work—a grim affair—but an exciting lark. The strong west wind and the sunlight pouring through broken clouds gave zest to our journey. White sandy beaches under somber green coastlines that would have been lost on a dark, overcast day extended tempting invitations. The red or Norway pine, whose color had been lost in the rain we experienced on Robinson Lake, now stood in full glory. Its tall, bare trunks with red-brown bark and its sparse graceful limbs gave color and character to many points and cliffs. The famous white pine, too, now stood at fine advantage. Sometimes it was mixed with red pine (its rather constant companion in the north woods); sometimes it had taken full possession of a point. I had first seen a virgin stand of this great tree on an island in Basswood Lake, where we started our journey. I had known this five-needle pine in the Far West, where it grows on high ridges—the last tree one sees as one passes into the arctic zone. There it is wind-blown and bent, twisted and gnarled. There it gains character and beauty from the adversity of the alpine ridges where it takes hold. In the northland the white pine has a different distinction. Here it is a stately queen of dignity and grace. Smooth bark and tall, straight trunks mark the tree. Its branches grow almost straight out, turning slightly upward at the tips. From a distance they seem to be tier upon tier, shaping up into a pagodalike structure. The early history of America could be written around this tree. No commodity was ever exploited more. It was the source of great industries and great fortunes.

Thoreau once wrote, "Think how stood the white pine tree on the shore of Chesuncook, its branches soughing with the four winds, and every individual needle trembling in the sunlight—think how it stands with it now—sold, perchance, to the New England Friction-Match Company!"

The white pine was a vital force in the American Revolution, starting with the edict of the king reserving the finest specimens for the Royal Navy. A revolutionary flag had the graceful white pine as its emblem. Our coins once carried it, making Thoreau say sarcastically, "Man coins a pine-tree shilling (as if to signify the pine's value to him)."

This early American history seemed to parade before me when I saw the virgin stand on Basswood Lake. And now that the sun shone at Sarah, lesser stands of the ancient monarch, whose golden pollen used to fill the eastern skies like a cloud, sent pleasant sensations up my spine. Here was a living emblem of America's early struggle for liberty, a tree so tall, so graceful, so majestic as to make its memory almost sacred.

We paddled around a point studded with white and red pine and beached our canoe in a warm cove lined with soft white sand. A pine snake lay dozing in the sun, and reluctantly left at our arrival. Dwarf bearberry—a midget compared with the lush bearberry of the Olympic Peninsula—grew thick above the sand. Behind it were a few paper (canoe) birch and a thick stand of black spruce towering eighty or a hundred feet. This tree—that used to supply the market with spruce gum—likes to have its feet near water. These black spruce are spirelike trees with stiff, flat branches. We started a fire with birchbark and made water for tea. Sig used a paddle as cutting board for the cheese and sausage. We ate our lunch in sunshine as warm as June's, while gentle waves washed the sand at our feet.

A granite cliff was too inviting to resist. Entering the forest, we worked our way around down timber, collecting beggar-ticks on our legs. Finding a series of ledges, we followed them to the top. Now we were on the north side, where lichens and moss are lush, covering portions of the granite like blankets. There were species almost without number. One piece of star moss was big enough for a bed and as familiar as the star moss flourishing on the high lakes of the Wallowas in Oregon. As I cleared the edge of the cliff and came to the top, I was on a thick carpet of gray-green caribou moss. Next to it was a patch of ground juniper in prostrate form, only a few feet high and fashioned into a circular clump. A red pine had fallen across the two. I chose it as a back rest and stretched out in the soft caribou moss. Sarah Lake was at my feet and the wilderness extended as far as I could see in three directions.

The struggle to keep this a wilderness had been long and painful. Sig Olson, who sat in silence at my side, was one of the stalwarts. There were many, starting with Teddy Roosevelt, who joined in the battle both here and in Canada. It was a struggle long in years and bitter in intensity. A vast acreage had to be protected against the lumberman. Private interests wanted to submerge these lakes for hydroelectric projects. These proposals had to be defeated. Roads had to be kept out so that moose, deer, and caribou could thrive in solitude, so that the lonely cry of the loon would not be broken by the automobile horn. The greatest struggle of all was to ban the airplanes. Amphibian planes could put down easily in almost any lake. They came by the dozens, returning men to their comfortable homes before dark of the same day. No more portages. No more hard struggles against a head wind. No more need to camp out under dripping trees. The airplane made this wilderness easily accessible. Record catches of fish promised to deplete the lakes. The roar of motors robbed this sanctuary of its quiet and solitude. So on December 17, 1949, Harry Truman signed an Executive Order that banned private flights by making the airspace below the altitude of 4,000 feet above sea level in this region an "airspace reservation." The courts sustained his action and over one million acres of wilderness regained the quiet and solitude it had known when the voyageurs traveled it in their birchbark canoes.

This afternoon on the cliff above Sarah Lake I relived every chapter in this historic struggle and I blessed the men and women who had saved this sanctuary for oncoming generations.

We had fair fishing on our canoe trip. Not the fishing of June and July, when the waters are alive with insects. This was almost October; May flies, black gnats, and all the other bugs that hatch in these waters had laid their eggs and died; the fish had gone deep; the frogs were silent. But Sig, who has made dozens upon dozens of trips into this wilderness area, knew every log, every rock, every point, every beaver house, every narrows where the big ones lay in wait. We used casting rods with spoons or plugs weighted with lead to comb the bottoms; and our rewards, though not munificent, were ample for our needs.

To the casual eye the waters in this chain of lakes seem alike. There is, however, a vast difference between them. The difference turns largely on the land they drain. Shafts of sunlight in some show a slight murkiness. These waters drain bog and muskeg. Others have a crystal-clearness. They drain highlands. Such are Robinson, McIntyre, and Elk, where the lake trout flourish. Other lakes, such as Crooked, which are darker, feature the bass and the pike.

The smallmouth bass is an old friend from New Hampshire, where I once spent my summers. And when the first one came to the canoe on Robinson Lake I was flooded with memories long forgotten, of the pools and sandbars of New England's waters. I scaled these Quetico bass as a storm broke over Robinson, a storm that drove us into the tent, for the heavens opened. When finally we cooked them over a cedar fire, they were as sweet and tender as any I ever ate.

We caught more northern pike than any others. These great fish, which reminded me of our New England pickerel, are not considered choice by many. But Sig made a chowder of them that was the tastiest I have known. In one pot he put diced potatoes and onions. He parboiled the pike in another pot so as to remove the bones and the skin. When the potatoes and onions were about done he added milk made from powder and the fish, letting the chowder simmer a while. Served piping-hot, this dish is my choice for the night when one is dead-tired and ready for the sleeping bag.

Steaks of the walleyed pike which we had in Crooked Lake and steaks of the lake trout which we ate at Elk Lake head the list. These fish are skinned and preferably fried or broiled. The walleye seems sweeter to the taste, the trout fatter and of finer texture. The preference is a close decision. Perhaps I chose the trout for aesthetic reasons. When the eight-pounder came to the canoe in Elk Lake, it flashed every color in the rainbow. That brilliance was soon to fade. But in the moments of excitement between the strike and the landing it had the splendor of coral seas.

This valiant fish is on its way to extermination in some parts of the northland. It was once a staple food taken in quantities from the Great Lakes. But the sea lamprey (which is shaped like an eel, though unrelated to it) invaded from the ocean, clinging to vessels that passed through locks and finally reaching Lake Superior. It has all but exterminated the trout from the Great Lakes. In the Quetico-Superior area they still cruise in safety and in splendor. The lamprey, which has no known enemy, takes as its sole nourishment the blood, body juice, and liquefied flesh of fish. Its mouth is a strong suction cup that brings the skin of the victim into contact with the armed tongue of the lamprey. The rocking motion of the tongue wears a feeding hole into the fish. The sea lamprey attacks lake trout as well as other fish.

Other bright memories of the canoe trip came flooding back on this last night of our journey. The portages had been vivid this fall. Those lined with aspen and birch were golden colonnades. Mountain ash showed crimson. The red maple was a plume of wine-red. There were red and yellow leaves underfoot, smooth granite rocks fringed with lichens, down logs covered with moss, small bogs fed by underground springs, rich stands of ferns. The shift from water to land gave more than relaxation from the confined positions in the canoe. It gave a closer view of the life of the north woods. Near one portage on Basswood was a large field of wild rice growing in shallow water. It was turning golden. Soon the Indians would come to reap it. One will paddle a canoe, a squaw sitting amidships. As the canoe passes gently through the rice, the squaw will use a stick to bring the heads over the canoe and then beat the kernels off. This precious cargo will be blanched in kettles with a fire under them, the rice being continuously stirred to prevent burning. Then will come the thrashing and winnowing, the product selling in the wholesale market at $1.50 a pound.

The botanical glories of the northland are not much in evidence in the fall. I had learned at the Wilderness Research Center that at least five thousand botanical species grow in this area. The director, Cliff Ahlgren, had shown me many of the prized specimens. Their heyday is spring and early summer. There were, however, some still in evidence. In addition to the tall purple and white asters already mentioned were extensive stands of the large white oxeye daisies. Once in a while a portage would show bunchberry dogwood in bloom in a protected spot. The partridgeberry showed red fruit as it draped itself over down logs and found footing in mossy ground.

At the end of one portage we put the canoe down in a stand of sweet gale, whose leaves are fragrant, whose waxy berries make sweet-smelling candles. And on many shores we found a large species of the horsetail from which whips were once woven and which the deer like when this plant first comes out of the water. Once we saw cranberries and, near them, the winterberry that seems to keep fresh all winter long. I saw two partridges or ruffed grouse on one portage, the male making a big fan of his tail and ruffing his collar. Where one portage climbs a sharp hill, a spruce grouse or fool hen went out ahead of us. We saw ravens, flickers, Canadian jays, pileated woodpeckers, the horned lark, kingfishers, white-throated sparrows. Some of these were on migration. But the woodpecker stays all winter; and so do the chickadees, who were common to most portages and who were singing their hearts out with their plaintive song.

Apart from the birds and an occasional red squirrel, the woods of the portages were quiet and still. Some portions were almost impenetrable, where down logs, berry bushes, and willow made a jungle. We had the canoe and four duffel bags to carry over each portage. Sig usually took the canoe, lifting it up bottomside down and resting it on his knees, taking the far side with his left hand, and raising it gracefully to his shoulders. I followed with packs. Often I sat on a granite rock by the trail's side, not to rest, but to feel the solitude of the north woods. They envelop one quickly and shut out the world completely. This was late September, when a deep silence settles over the woods. Not many birds were singing, for this was not the time of mating and home building. Even the frogs and swamp crickets were silent. When fall arrives, one can sit for hours in these north woods and almost hear one's own heartbeat. Their loneliness reminded me of the voyageurs who first cleared the paths for these portages and negotiated them in a dogtrot. These voyageurs, whose songs broke the silence of the northland, must have found beauty and glory in the deep shade of pine and spruce and birch. They too knew the star and the caribou mosses and saw infinite beauty in the delicate construction of the lichens. The colored leaves underfoot, the scatterings of asters and daisies, a flash of dogwood, a stand of black currants, and the gold and red trees are enough to set anyone singing. Once, along a portage to Elk Lake, I found an old acquaintance. A stand of alder decorated a granite knoll—alder of the same species that I know on the Olympic Beach. Only this was dwarfed and small, but nonetheless a welcome friend.

These portages bring discoveries of other kinds. Each was to me an exploration, a launching into the unknown. And the same was true of Sig, though he knew each portage intimately. Each is filled with expectancy, whether it be short or long, steep or flat, winding or straight. Somewhere ahead is a new sapphire lake with a personality of its own. The first sight of it through the trees is always a joy. It stirs the imagination, beckons one onward. It promises the prospect of beavers slapping the water with their tails, loons calling, mergansers skimming the water, graceful points painted red and gold, a lake trout flashing color in the depths, moose feeding on lily pads in a bog. Every lake is a new chapter, with its own theme, its own challenge. Portages bring the expectancy of high adventure ahead. Every streak of blue seen through stands of pine and spruce of the northland has the magnetism the voyageurs felt. It invites one on and on—on to flaming sunsets, moonlight nights, and majestic white pines murmuring in the wind.

There are bear in the north woods; and every night when we made camp on the mainland we collected our kitchen and pantry and put all the supplies in easy reach of the tent. But we had no marauders. There are timber wolves and lynx in the Quetico, but we saw no sign of them. Though the great migration of birds had already taken place, the ducks were still to come. Only a few black mallards were present, and some teal. But once when we rounded a sharp point on Sarah Lake, a flock of bluebills went out from under our bow with a roar of jet planes. There were a dozen in the flock, but they flew as one, barely skimming the water for a half-mile and then whirling in unison to disappear over a stand of red pine that adorned the next point.

We received many scoldings from the red squirrels that frequent the Quetico. But the severest of all was when we finished a portage through birch, maple, hazel, and jack pine at Sarah Lake. The squirrels, who were very active in the jack pines, seemed possessed at our presence.

This two-needle pine that we use for lodgepoles in the Far West has many legends attached to it in the northland. It is supposed to make women who come close to it sterile, to poison the soil, to bring ruin to livestock. Its magic is so potent that the voyageur would not cut it down.

This jack pine is a stubborn, knotty, stunted tree that thrives under adversity, and dies not much later than a hundred. Though good only for lodgepoles and pulp, it does the northland great service. It covers thousands of square miles of cold, sterile ground where nothing else as tall might grow. And it performs heroic service in the cause of conservation when forest fires come. The jack pine is jealous about its seeds, keeping them in the cones for twenty years or more. But when great heat comes, as in a forest fire, the seeds are discharged. Slash fires, like crown fires, can be very hot and destroy the seeds. But when extreme temperatures are not created by the holocaust, the seed often escapes destruction, even when the parent tree is killed. Green cones are good insulators and are not highly inflammable. That is why seed in cones often lives through forest fires and falls on the freshly burned surface to germinate. That is why the jack pine comes in first after a fire, serving as ground cover when the risk of depletion is greatest. And when June comes, the waters of Sarah Lake will be covered by its pollen.

Perhaps a fire was responsible for the thick stand of jack pine on the edge of Sarah Lake. Perhaps the red squirrel was the agent. They seem to know when the cones of the white and red pine are ready to explode. They always seem to get there first and obtain their own supply. They can work more leisurely on the jack pine. But whatever the pine, the squirrels are ceaseless collectors of its seeds; and they store them in countless places. They seem to have a memory that is not too good. For many of their caches of pine seed go untouched and survive to start seedlings on their way to a new forest.

One day on Brent Lake we had lunch in a lovely grove of red pine. We used for fuel some dry pussy willow that is locally called diamond willow. It is a reddish wood with diamond-shaped knots—a hard, hard wood that the northerners like to whittle. And it makes a bright fire. After lunch we fished several beaver houses and logs protuding from the shore, landing a few northern pike. We had put up our rods and traveled a few hundred yards when a mother otter surfaced on the starboard. She had nothing but vituperation for us; every utterance was a hiss as she glowered at us, her eyes dancing with anger. These otter are the fastest-swimming mammal in these waters, able to overtake any trout. She would disappear, only to reappear fifty yards on the opposite side of the canoe. In a jiffy she would pop up in front of us, then behind us. The game—which was to conceal her young—kept up for a half-hour or more. Sig and I, sure that the young ones were beached somewhere on the rocks, searched up and down for them. The picture prize, had we made the discovery, would have been invaluable. But mother planned well; we did not find her babies. And as we headed west, she came almost halfway out of the water to give us a farewell hiss.

The night we camped at McIntyre was windy, blustery, and chill. The desirable campsites with polished granite beaches were exposed to the wind. They would have been ideal when the black gnats and mosquitoes were abroad. But tonight we wanted shelter. We finally found a secluded spot in a grove of red pine that was thick with needles, and flat. No one had ever camped here before. We built a fireplace, gathered wood, and pitched our tent. The sun was low, and supper was cooking, when a sea gull suddenly put down a respectable distance from us and sat there, filled with expectancy. Suddenly a loon swam by the point. Loons have four calls: (1) the "talking" call; (2) the "wail"; (3) the "yodel"; and (4) the "tremolo," or "laughing," call. The calls we had heard on previous nights were the "yodel" calls. This loon on McIntyre Lake was making the "talking" call. He must have been talking to himself, for no other loons were in view. His were single-chatter, one-syllable notes uttered in a calm voice—the kind of talk old cronies sitting in a canoe might exchange. Suddenly he changed his call. He had seen us and was excited. He picked up speed and sounded the alarm with the "tremolo" call. This call announces danger—the presence of an intruder. It expresses worry and concern. It was a high-pitched "tremolo," the half-dozen notes coming fast and with precision. Our friend did not cease calling until he was far away and out of sight.

These days in a canoe brought life in a new dimension. In calm waters we seemed to glide somewhere between earth and heaven, silently and gently. When the wind picked up, or changed its course, the canoe seemed to stiffen like a spirited horse, only to relax when we established a new rhythm with our paddles. As our strokes settled to a steady beat, the canoe once more bent to our will, and paddling became a routine, effortless exercise. But when the big blows came, the contest between man and water became intense. Then it seemed we were in an elemental battle that called on every bit of our ingenuity and most of our strength.

There was a powerful wind the morning we headed west across Brent Lake. We felt the full force of it as we bowed to it and leaned on the paddles. Sig, in the stern, sought wherever possible the lee of a point or an island. That often took us on a zigzag course while crossing broad open reaches as we tacked for another calm cove. Once we reached quiet water, we would let the canoe drift under granite cliffs bedecked with lichens or sit in silence under majestic white pine. Our strength renewed, we would nose out around another point and meet the fury of the wind head on. For one stretch of a mile or more we had no protection; the full force of the wind held us fast. I do not know how many strokes we made without moving. My mark on the shore was a red pine. It showed us stationary for about two minutes. Then once more we moved forward—inches at first and then with a steady drive. And as we moved, Sig burst forth in song: "Back home again in Indiana."

Singing while paddling was the manner of the voyageurs:

> Row brothers, row, the stream runs fast,
> The rapids are near and the daylight's past.

There was the voyageur song about Thoda, who spurned three barons for a youth she loved:

> Oh, my heart so true
> Is not for you,
> Nor for any of high degree.
> I have pledged my truth
> To an honest youth,
> With a beard so comely to see.
> Oh, the violet, white and blue.

Another favorite was *On, Roll On, My Ball I Roll On:*

> Way back at home there is a pond,
> On, roll on, my ball, on.
> Way back at home there is a pond,
> On, roll on, my ball, on.
> Three bonny ducks go swimming round,
> Roll on, my ball, my ball I roll on,
> On, roll on, my ball I roll on,
> On, roll on, my ball, on.

These voyageurs were the heroes of the early days, not Alexander Mackenzie, David Thompson, Alexander Henry, and the others who wrote the chronicles. Those authors were passengers in birchbark canoes paddled by the voyageurs, who sang their way to Winnipeg. (The younger Henry, however, set the solo record for travel by birchbark canoe between Pigeon River and Fort Francis, at eight and a half days.) The paddlers received rations of rum and corn—that is all. Meat and fish of course went into the pot, including bear meat, which they liked because all fresh meat was at a premium. But there was precious little time for hunting and fishing. They traveled until after dark and were on their way by two o'clock in the morning. Their baggage was in ninety-pound units; and each man carried two of those packages on the portages. Their birchbark canoes were not too tight. Hard blows, high waves, or fast currents would cause them to leak. The voyageurs carried pitch pots with them and often beached for repairs, lighting a fire and melting the resin.

On and on they went, hour after hour, day after day, singing as they paddled, singing the whole day long. These French were curious people. The Spanish and British came to conquer. The French came more reverently. They passed through, leaving some marks—portages that they cleared and high trees that they trimmed for landmarks. Some suffered torture to convert a few Indians; but they were not torturers in turn. They left musical names behind, but no monuments. Their presence did no more damage than the passage of their rhythmic paddles through the water.

When we were in Brent and Crooked we were on their highways. And the cadence of our strokes made me feel I knew them better. They too loved the feel of the canoe in water, the sound of loons, the splashing of moose in bogs, the whir of mergansers going out from under the bow and turning at right angles to give one the broadside view. They loved the mysterious shape of woods in morning mist, dancing moonlight on calm waters, the roar of the falls ahead.

They were very much on my mind this last night of our trip, when we camped on the tiny island below the Lower Falls. I sat into the night until the mist had enveloped the falls, wiping them from sight. I went to bed, only to get up again to revisit the moonlight and to see the mist embrace the entire island. When I finally retired, I lay for a while, listening. Above the roar of the falls I seemed to hear voices—snatches of old French songs, a shout, a command, indistinct words and phrases, laughter, the scraping of canoes on granite, heavy packs hitting the ground. Those were the voyageurs who had passed this way two hundred years before my arrival. Now I was with them as they finished their portage and headed their canoes down Crooked Lake.

THE PASSING OF THRUSHES

Tom Anderson

A cool mist was in the air. It didn't seem to fall, instead it floated, suspended droplets resisting the pull of gravity. The sodden skies over Lake Superior encouraged us to stay indoors and relax in the quietness of the cloud that surrounded all.

Beyond the window, flashes of flittering greens betrayed the late summer restlessness of warblers and thrushes as they worked nearby thickets for the promise of insects. Their days in the northwoods are few as they begin their southerly trek.

A sudden thump against the windowpane seemed out of place for such a quiet afternoon. It was the unmistakable sound of feather and body hurtling against the resistance of glass. By the sound of the impact, this was no fleeting brush against the pane—it was solid.

Immediately, I rose from my chair and peered out the window, hoping I would see only a lawn of wet grass. Tiny feathers were still drifting down with the mist as I spied a bird, slightly smaller than a robin, lying on its back, wings spread and legs sticking up, stiffly askew.

I stepped outdoors and quickly scooped up the bird. Upon picking it up I realized that the bird was a thrush. The buffy eye ring and lores gave the bird a bespectacled appearance. The "eye glass" markings also confirmed the bird's identification as a Swainson's thrush.

In the early 1800s, a British zoologist named William Swainson, like so many European scientists, was drawn to the North American wilds. For a time, he and well-known artist-naturalist and ornithologist John James Audubon explored and collected together.

Swainson was commemorated for his work in the New World by the naming of three bird species after him: Swainson's hawk, Swainson's warbler and the Swainson's thrush.

Though this bird is by no means spectacular in plumage, it makes up for those shortfalls with its melodious song. It would be among the best of honors to have one's surname associated with such music.

The song is a flutelike spiraling climb of notes that are most often heard as the day's shadows merge together. In the more southerly deciduous forests in the central part of the state near my home, I often pause on a summer evening to listen to another thrush, the veery. Its call is a similar spiraling of notes, except they descend rather than rise. I remember the Swainson's symphony as a "backwards veery song."

Some of my fondest memories in the northern canoe country include those moments at dusk when the stilled waters of the lake reflect the sharp spires of spruce and fir and nearby the Swainson thrush claims its piece of boreal domain.

In my cupped hand, I carried the fallen bird into the house. It seemed near death with its eyelids half closed and its beak spasmodically opening and closing as if it were gasping. The left leg was folded close to its body, but the right leg was rigid.

Indoors, the bird was the center of sympathetic attention. My two young daughters were curious. Britta, aged four, was shy about touching a barely alive bird but more bold in her questions. I explained that the bird, a cousin to the robin, had accidentally flown into the window. Perhaps it had seen the reflection of a misty, sullen sky in the glass and headed for it.

I pointed out the fleshy, yellow gape, the corner of the bird's mouth, which indicated that this bird thrush was a young bird born only two or three months ago.

Britta's eighteen-month-old sister, Maren, squealed a semblance of "bird!" and brashly poked her finger into the bird's breast.

Realizing that the bird was under enough stress without the thrusts of curious fingers, I placed the bird on its back in a cardboard box. The dark box would hopefully relax the bird.

I had considered that perhaps the most humane thing to do would be to squeeze the bird's rib cage, killing it quickly and quietly. But there was too much shine in the bird's eyes. I decided to keep the bird in the box for a while to see if it might recover.

Back indoors, out of the weather, we resumed our visiting and all was well.

An hour later, Britta and I checked on the bird. Quietly I opened the box and as the gray light of day shone in, a fully recovered thrush rocketed out, disappearing into the woods bordering the yard. It was reassuring to know that this thrush was given another chance to make its nocturnal autumn flights to its winter haunts in southern Mexico or South America.

Two days later, some thirty miles north, we sat in a cabin enjoying a leisurely breakfast. Suddenly we were interrupted by another thud against the window. Again we found a bird, and, coincidentally, it was another Swainson's thrush. Like the first bird, this one also had a yellow gape proving its youth. However, this bird had no sparkle to its eyes.

Britta had no qualms about holding this stilled bird, for it was dead and, therefore, posed no threat. For half an hour she sat with the bird cradled in her hands. Later we would bury it.

And so for one Swainson's thrush, the restlessness of fall still prods it mysteriously south. For another thrush, there will be no southern jungles, and the passing of summer is a lifetime.

THE LAND IS ALIVE
WITH WOLVES

Jim dale Vickery

This truth stuck to my mind like a wet tongue on cold steel as I crouched down to feel fresh timber wolf tracks with my bare fingers. It was a winter afternoon, and I had just finished working in my writing cabin on a granite ridge near the heart of Superior National Forest's wolf country. But the tracks, made beneath blue skies, nevertheless came as a surprise. Suddenly my sluggishness was gone and I was electrified—as if plugged into the sunshine itself. I glanced around, found fresh deer tracks among the wolf tracks, then noticed a snow-covered bush stained yellow with wolf urine near the cutoff to the main cabin, where Chris Tröst was waiting for me.

I was not—nor had I recently been—alone.

The main cabin across the cove would have to wait.

Soon I was backtracking the wolves; I was a tall, relatively gangly creature shuffling through ash and birch swamp down a trail up which wolves had come. Later, on the lake at sunset, I would see that four wolves had been my way. But no matter the number. For the moment I was obsessed with the snowy, five-inch-long tracks. Tracks leading off the trail into brush. Other tracks coming through alders to join the main path. The wolves had been sniffing things out, I supposed. Taut for the flush. Hoping to feed themselves, according to an old Russian proverb, by their feet. Maybe on snowshoe hare. Maybe on beaver or grouse. But most likely on white-tailed deer here in northeastern Minnesota.

With, I had to add, an occasional dog.

Like Bojo.

Just when I was ready to end my wolf tracking for the day I found greenish wolf scat—embedded with bone chips and hair—on the trail in front of me. The connection between wolf scat and my husky lunged up at me like a shadow on a moonlit night. For the moment I was stunned: hit in the chest. I was trapped in the steel jaws of a time I couldn't shake: the time Bojo was killed, when my encounters with timber wolves made a quantum leap forward and left me a changed man.

I suppose my run-in with wolves began when I moved to the North Arm of Burntside Lake near Ely, Minnesota, in May 1979. What I wanted, like many newcomers to the region, was wildness: to live near the Boundary Waters Canoe Area Wilderness in Superior National Forest, locus of three million acres of trees, lakes, rivers, streams, bear, deer, moose, and wolves. Across the border, in Ontario, was Quetico Provincial Park, another million acres of wild terrain. Precambrian, Canadian Shield stuff: a vast plateau of billion-year-old granite extending southwest of Hudson Bay. It is a land Ojibway knew, and Sioux before them—a place where they left red ochre pictographs, bound with fish oil and mystery, on rock faces fronting canoe routes. Here French Canadian voyageurs paddled fur trade highways in the 1700s and 1800s.

Bordering this Quetico–Superior arena of history and space, watching over it from the north, was the Ontario town of Atikokan ("caribou bones"), beyond which stretched incomprehensible wilderness, boreal and subarctic, panning out into the great Arctic itself. There were only three east-west roads between the Arctic and where I came to lay my head.

Lindskogs Resort was the name of the place, a pine ponderosa of a couple hundred acres sandwiched between Burntside Lake on the east and the Boundary Waters' Cummings Lake country on the west. Ely, then a mining and tourist town of 5,000, lay nine miles southeast as the crow flies or the ski glides; twice that distance by road. Although the resort was a bustling place of rented cabins and second homes in summer, in winter the wildness came home.

The change came around November, when gray skies and freezing temperatures brought ice to coves and shoreline rocks. Most canoeists were gone by then, as were swimmers, fishermen, and hikers. Only orange-jacketed deer hunters remained. After deer season the country was left to those who wanted it.

Like me. My first two winters were spent alone at Lindskogs, my nearest neighbor a toss-up between a silver-haired Swede in a log cabin on a ridgetop and a radiologist a mile up the North Arm Road. I lived in a one-room log cabin the first winter, in a two-room frame cabin the next. Around me, statisticians claimed, were 800 to 1,200 wolves in a five-million-acre area. Here was the last viable population of timber wolves in the United States outside of Alaska.

True, there were a few wolves in the Rockies of northwestern Montana, twenty-six in northern Wisconsin, another two dozen on Isle Royale, Michigan, and rumors of a wolf comeback in the northernmost eastern states. Yet only in northeastern Minnesota between Lake of the Woods and Lake Superior did wolves maintain a healthy foothold on American land and in the American mind. It was in the latter they had caused such a ruckus in the 1960s and 1970s. As wolf populations fell to the American hunter and trapper, a controversial outcry by environmentalists led to a blast of legislation designed to protect the last hangouts of *Canis lupus* in the Lower 48. Bounties on wolves—which had baited both casual and professional hunters for decades—were stopped in 1965. Then, in 1970, authorities closed Superior National Forest to the taking of wolves.

The final blow, as some might look at it, came in 1974— one year after passage of the Endangered Species Act—when the taking of all wolves in the forty-eight coterminous states was forbidden. This, five years before I settled in at Lindskogs.

I had never seen a timber wolf before moving to Burnt-side Lake. Never, despite being a lifelong native of northern Minnesota. I grew up in Red Lake Falls and Crookston in the northwestern part of the state, in the Red River Valley, a region with the same latitude as Ely. But you don't see wolves in Crookston country. You see farmers, sugar beets, and red foxes.

You don't see wolves, either, in White Earth State Forest and Indian Reservation in north-central Minnesota. While living there for most of four years in the mid-1970s, in a primitive cabin a dozen miles southwest of the Mississippi River's headwaters, I never in my daily walks found tracks or other sign of timber wolves. Brush wolf or coyote, yes. I found one that had been shot hanging by its neck in a tree along a logging road, and I sometimes heard them yipping and howling. But varmints didn't last here much past their first or second sighting.

Not until 1977 did my tracks finally cross those of timber wolves. While on a canoe trip through, and beyond, northern Minnesota's Voyageurs National Park, I found fresh wolf tracks trailing moose tracks on a sand beach of Lilac Lake. The next day a friend and I heard a wolf howl from nearby ridges.

The following spring, during a snowshoe camping trip in the Boundary Waters Canoe Area with wildlife artist Dan Metz, we found signs of a wolf kill below Devil's Cascade: vertebrae, deer fur, and bloodstains on snow. There, at a small pool of open water, wolves had attacked a deer that had come down to drink. Parts of the same deer were found farther down the Little Indian Sioux River where it widens into Loon Lake. Again the next day, near the end of the snowshoe hike to Lynx Lake, Metz and I found fresh wolf tracks cupping gold sunlight from a setting March sun.

Canis lupus itself, however, had scarcely crossed my life.

This changed at Marshall Lindskog's ponderosa, where there was more wolf traffic than I had anticipated. I often heard wolves howling at night as I walked down to a hole in the lake ice to get water. Or when I stepped outside in subzero air to fetch an armload of firewood. I saw wolf tracks while cross-country skiing and snowshoeing, and found it curious when fresh wolf tracks followed mine back from my afternoon jaunts.

I tracked. And I was tracked.

One morning I found wolf tracks circling the cabin and on the path to the outhouse. Alongside the path were depressions in the snow among the alders, indicating I hadn't slept alone during the night. Urine yellowed the snow beneath several white pines.

Was I being stalked? Watched? Played with like some wolf scientist in a Farley Mowat book? I never cried wolf. I kept the excitement to myself.

More soon came. On a blustery night in midwinter, while talking with a friend about shelters, I heard an odd noise outside the cabin window. I had left one of Marshall's burbot carcasses on a woodpile for the birds, and the strange sounds made me think a flying squirrel or pine marten was doing some nocturnal dining. When the subject of tipis came up with my friend, I mentioned a book I had; soon I was breaking trail through the dark to my summer cabin, where most of my books were cached until spring. Halfway to the cabin, as I sank knee-deep in blizzard-driven snow, I spotted canine tracks.

My first reaction—rooted in a boyhood spent in towns—was *dog*. But it's hard to tell track size in deep snow: As soon as the leg is raised out of the snow, fresh powder slips into the track.

I entered the cold cabin, found the book, and headed back to my place on a different route.

The driveway had less snow, but here too were tracks, unmistakably wolf in size. They were going my way. My hackles rose as I stared at the tracks illumined with my flashlight, the child in me—no, the wild in me—coming alive with ancient reactions in a universe of dark. The tracks sometimes left the road to meander among brush, yet always returned. I wondered what scent, what hunger, possessed this track maker as I traced the tracks to the back of the cabin. To the woodpile near the window. To the burbot carcass now pulled to the ground.

A new hunger dogged me as I entered the cabin. I wanted to *see* a wolf. I was getting weary of reading signs and guessing scenes. Perhaps I should sit up some moonlit night awake to motion among trees and brush outside the window. Bait wolves with burbot, if I had to.

But, *no*, I told myself. That would be like sitting in wait for an eagle. That would be too much like holding one's breath in hopes of a vision.

Bojo was the key to wolves for me. It was she—a purebred Siberian husky—that sharpened my senses and quickened my quest to get closer to the mystery that is *Canis lupus*.

I bought Bojo for $100 when she was six weeks old in April 1980, from friends at the *Ely Echo*, a weekly newspaper where I worked part-time for two years. I had always wanted a dog, and in northern Minnesota none is more at home in the cold than the husky. The Siberian's pelage, with its white eyebrows, white cheeks and chest, and other light markings, and its curved tail give it an added beauty—like the wolf's—that sparks the soul.

Too, huskies are rooted historically to the Quetico–Superior. Indians had used huskies for centuries to pull food and fur across the island-pocked snowy expanses of today's Minnesota–Ontario border, even as far south as the southern shore of Lake Superior. There Ojibway of the Apostle Islands once greeted eighteenth-century Frenchmen with "*Bojo . . . Bojo*"—a broken version of the salutary French *Bonjour*.

Bojo. The name not only appealed to my French heritage, it also anchored time and husky to the land. I spoke the word countless times, rubbing it into my pup's mind like a salve.

Part of my attachment to Bojo was obviously emotional. Shortly before I bought her I was divorced from my first wife; call it incompatibility. The sorrow and confusion were compounded by another relationship in Ely which left me stressed, ripe for a much-needed return to the simplicity and peace for which I had moved to the north woods in the first place. Bojo refocused my attention. We walked together daily. She made me chuckle in my solitude as she skidded in circles on linoleum flooring in the morning when she judged day had come. We played tag on old logging roads, went fishing and berry picking together, and once encountered an adult black bear which Bojo harried before luring it toward me.

Sometimes when I told Bojo she was the best dog in the world, I ended the flattery by howling: first a little yip and a whine to get her attention, then a deep throaty howl coaxing her to join me.

Sorry was the world's breaking heart when Bojo sang. Woe was the wildness calling, calling somewhere behind her blue eyes.

The last day of August I took Bojo for a walk down a trail we had walked a dozen times into the Boundary Waters wilderness. It was a private path known to few, a mile of forest and flower leading to a small lake fringed with pines. There, on snowshoes, I had often "checked esthetic traplines" in search of winter's wonder. There, in summer, I had frequently fished, always returning to my cabin with a stringer full of beauty. It was a sunny Sunday afternoon as Bojo and I nosed through brush. I didn't think twice as she broke off from a trot in front of me to hunt down a scent. This was her wild haven, certainly the nearest thing to paradise a husky could find.

The bushes closed behind her. Without a glance backward she was gone.

I was putting a Lazy Ike on my fishing line fifteen minutes later when I heard Bojo scream. It was an incredible cry: high, sharp, earnest, terrible. It wasn't the ferocious snarling and barking of a fight but a bawling howl of tearing pain. I ran down the boulder shore with my fishing rod in my hand, then back up the trail toward Bojo.

Her cries stopped. I whistled and called. Silence. Not a leaf stirred.

Hearing nothing more, and feeling helpless, I turned around toward the lake to resume fishing. I fought worry. Puzzled? Yes. But it was broad daylight in late summer a mile from my cabin. Familiar ground. My backyard.

I didn't get twenty steps when Bojo started screaming again.

Again I ran down the trail calling for her. If only I could have heard a branch crack or seen something move. I was desperate. Anxious. *Where in the hell is she?* I wondered. *What's going on here?*

Bojo's howling, her yelping, stopped.

I looked for her for the next five hours, then headed home, where I hoped she had returned. No dice. Her food, her toys in the yard, even the old leather lineman's glove with which we had played tug-of-war, lay untouched. I recalled her cries while I lay in bed. I had been her only companion. Her alpha male. It was *me* she had been calling for. And I had been helpless.

I was back in the woods at dawn the next day. I searched through bogs, up ridges, through thick fir windfalls, around and across swamps. I looked, called, and listened, death drawing nearer one step at a time. On Tuesday, friend Dan Kuhl and I scoured the woods with his dog. Near where I had last heard Bojo we found a pile of dark, diarrheic scat. I stirred it with a stick, found hair, and examined it closely. The hair was tan and white. Bojo's colors. The mystery of her fate was solved.

But what, exactly, had killed her?

Any noise on Bojo's fateful afternoon in those windless woods would have been heard, so I ruled out a bear. Bears walk more like moose than deer in brush; fearful of little, they muscle their way through obstructions oblivious to the noise. Besides, Bojo would have barked, as I had heard her do at bears before. A fisher *might* have killed Bojo, but she was already almost twice a fisher's adult weight. Finally, Bojo could have fended for herself against pine marten, bobcat, or lynx; at least I would have heard both foes in the fight.

What killed Bojo moved silently through the woods, killed quickly, and ate dogs.

Almost every woodsman I knew cried wolf. Particularly Marshall Lindskog. He had hunted deer, moose, polar bears, grizzlies, and wolves in his day, and he was confident a wolf had killed Bojo. The dark cow-pie scat, he said, indicated much blood in the predator's digestive system, possibly gulped raw meat. All I had to read was an inkblot of sounds, scat, and reconstructions of the scene. Yet two events seemed to close the case as far as I was concerned.

Call it primary evidence. Notch it up as wolf sign.

About two months after Bojo was killed, I was hunting ruffed grouse in Ole Lake country, where Bojo and I had hiked and occasionally camped. Beyond Ole, the trail led up to a high granite ridge skirting the east shore of Silaca Lake before dropping down to an old logging road connecting Coxey Pond and Burntside Lake. As I walked east, homeward, with the setting sun at my back, my eye caught something moving in the roadside brush. The wolf was passing in front of me on a game trail from right to left, about thirty feet away, and as soon as it reached the road its identity was certain.

The wolf trotted for one golden instant through a sunny spot of needle duff on the road. Its size, grace, and silence, its utter smoothness and confidence of motion, stoned me in midstride. My .22 rifle lay on my shoulder. Never had I been more riveted to the immediate moment. I was locked in a wilderness saunterer's satori.

Yet my mind made the Bojo connection: *This wolf could be her assassin.* We weren't more than five miles from where Bojo had been killed, a distance a healthy wolf can cover in an hour. This wolf, moreover, was coming from the general direction of Bojo's last stand. Perhaps it was making the ridgetop rounds of its territory, moving through the woods like a salmon in a stream of scents and false starts. Or perhaps it was just sniffing out deer yards in a buffer zone between wolf packs. This I could never know. The truth was furred and on the move.

The wolf picked up its trot through a stand of jack pine until it was out of sight. I stood silent, listening to its fading steps patter like raindrops on dry leaves. A *wolf.* Not for a second had I thought of firing a shot. I felt sensually heightened and transformed: astonishingly alive in a land that gave as well as took away.

My encounters with timber wolves, and the enigma of Bojo's death, assumed a new dimension when Bud Lindskog, Marshall's brother, shot a moose off the Echo Trail near the Little Indian Sioux River. After butchering the moose, Bud ditched the large bull's head at an old dump along the gravel driveway into the resort. Soon there were large canine tracks at the dump. The nose, ears, skin, and flesh of the moose head went quickly. One day I found diarrheic scat among the tracks. By then the skull had been tugged apart, its seventeen-inch lower jaws—stained with dried blood—scattered on a green bed of club moss.

The scat was exactly like the dark, soft stuff in which I had found Bojo's hair. I stirred this new clue with a stick, wolf sign spinning kaleidoscopelike in my mind on the cutting edge of fundamental Canidae knowledge.

The clincher came a day later, when Chris and I were standing on the shore of Burntside Lake's North Arm. Next to us—not more than a thousand feet from what remained of the moose skull—was a neighbor's log cabin. Suddenly we heard a scratching, shuffling sound as if someone were crawling out from beneath the cabin. When it had cleared the last porch log the wolf stood, twenty feet away, looking straight at us, its reddish-brown coat mottled with black. Then it loped off into the woods. Before disappearing it stopped to look at us over its right shoulder.

"Wolf," I muttered to Chris in the shock of recognition. But with the word said the wolf was gone.

Wolves seemed to be crawling out of the woodwork in much of Superior National Forest in 1981–82. In a bizarre incident, Ronald Poyirer—his clothes saturated with buck scent—was knocked down by a wolf in a spruce swamp near Brimson, Minnesota; after a scuffle the wolf took off, leaving Poyirer with a few scratches. Several leashed dogs were killed near Babbitt. More were killed near Section 30 southeast of Ely. Some dogs, left free to roam, simply disappeared, their names last seen in the want ads of local newspapers. I winced every time I heard of a killed dog, but still, I couldn't hold it against the wolf.

It came down to this: I couldn't begrudge wolves their desire for meat. Food: A wild wolf needs anywhere from four to ten pounds of meat a day, and I couldn't find their procurement of this any more repulsive than a pine marten's killing a squirrel, a gray owl's killing a grouse, a grizzly's killing a caribou calf, or a husky's killing a mouse. Bojo, like a fox, had been a great mouser. For me to try to teach her not to spring through tall roadside grass while hunting mice would have been like trying to teach a bass not to swim.

Clearly, there was a circle going on. Bojo had stumbled into it, and I had witnessed it. To draw near it, and to feel unto pain the life-death dance, was one of the primary reasons I had chosen to live in the shadow of the wolf.

This cycle, moreover, is what I now suspect most great wildlife artists tap in their best paintings of wolves. The wolves therein are always alert, often moving, on the run, one ear cocked for prey, the other swiveling for sign of man. They are hunting, feeding, thirsting for banquet. Their wolves hunger and seek; full of energy and strength, they move from darkness into light, from obscurity into clarity. The beauty here lies in truth. It is cyclic. And it is dramatic.

As for Bojo, would I have shot the wolf had I come upon the killing with a gun?

Perhaps. Despite the threat of a $20,000 fine. Despite a possible jail sentence of five years. Despite even a strong fondness for the wolf and the wildness it represents. Some protective instinct of my own might have pointed the gun at the wolf and pulled the trigger. My reaction might have been the same knee-jerk contortion with which a cattleman protects cows. Call it the shepherd instinct.

On the other hand, my armed presence would likely have spooked the wolf off, leaving me mouth agape and Bojo with a good lesson. I'll never know.

It's something to think about, though, as I walk these woods and canoe these waters, and especially when sitting around campfires listening to wolves howl. When wolves speak, the night speaks: It seems to be a language understood intuitively by all who hear it.

I heard that language once again last New Year's Eve. I stepped outside shortly after nine o'clock to watch the moon rise orange and two-thirds full above a serrated forest horizon. Stars whitewashed the sky at ten below. A dull concave lens of northern lights hung over Pine Island, similar to—but lighter than—the auroral glow the night before.

For the joy of it I started to howl toward the moon, my cry carrying across the lake cove below into last year's last darkness. I howled again, then several times more. I expected nothing. Yet when I turned to go back into the cabin I heard a close reply. It came from the east, out of the silence, a wild voice rending the night. A wolf was howling back to me. Then another. Suddenly there were wolves howling out on the lake past Cedar Point, and to the right of me in Rainbow Ridge country. A pack had split up and surrounded me. The last howls rolled off Rainbow Ridge as the risen moon threw a dull yellow light into the night. This on New Year's Eve.

Twice during the night as I waited for midnight to strike I stepped outside because I heard quick, excited howling and barking south of the cabin. I found wolf tracks on our snowshoe trail in the morning. Another set of tracks—these by a loner—crossed the cove beneath the cabin. The wolf had been running, kicking up wads of snow crust, and although I hadn't actually seen it I nevertheless visualized moonlight glinting off guard hairs as the great wolf, driven curious by my wails, policed its range.

Despite my five years in wolf country, I still hadn't seen a wolf-deer encounter by the time I moved to a new cabin southeast of Basswood Lake in 1984. This wasn't surprising. Many people who have lived their entire lives in northern Minnesota have never see a wolf. During a period of twenty years, wolf biologist L. David Mech saw only about a dozen wolves that were not tracked down by radio collar or seen from a plane. What wolves remain in the contiguous United States—where they have been trapped, snared, shot, chased to death by people on snowmobiles, and poisoned—are around today because they have made it their business to steer clear of man.

My own experiences with wolves had obviously been by happenstance. Only on such terms—if lucky—could I expect more. It happened on a late winter day two years ago. I was snowshoeing home on a bright afternoon after escorting friends to their car a mile from my cabin. As I topped a small rise, I noticed something moving on the snow-covered ice near an island's point out in the bay. There was no mistaking the wolf as it loped easily along the island shore, paralleling it, its head low, hunting scents and tracks, then raised to scope horizons. The wolf's coat, more brown than the normal gray, looked brushed and full.

I ran toward the cabin as soon as the wolf was out of sight. I knew Chris was inside reading, no doubt sitting by one of the cabin's six lake-facing windows, but it wasn't likely she'd be looking outside. I ducked behind a rock ledge hoping the wolf wouldn't see or hear me. I was at the cabin door in a moment, out of breath, bending over, unstrapping the snowshoe bindings, and trying to talk to Chris at the same time.

"Wolf!" I said between gasps. "Grab your parka and follow me to the top of the ridge behind the cabin. We'll have a lake view in three directions."

We watched two wolves come together in a narrows between the lake's two main islands. One wolf had cut across an island, perhaps to flush out a deer; the other had gone ahead to round the island's west point. They met in the narrows as we watched them with binoculars. What they hadn't seen yet was the deer across the lake to our left.

The deer was alone on the lake ice next to shore, its white flag up, its gaze over the left shoulder toward the wolves. It raised each leg slowly as it stepped forward. Only anxiously did it look away from the wolves to see where it was going. Chris and I wondered—binoculars at eyes—if the wolves, now out of sight on an island, would pick up the deer's scent. We also wondered if we were about to view a wolf–deer chase complete with the tearing, choking kill, the ripping of meat and cartilage to bellowing whitetail cries. a fearsome eruption of violence on a clear, cold day, a clash of tooth and muscle.

It had been watched before, this killing. It had been seen by biologists, woodsmen, and the occasional canoeist—like the late Quetico–Superior naturalist Sigurd Olson, who eddied into it by chance in the 1920s.

Sig Olson had been paddling down the Basswood River on the way to Crooked Lake during his guiding days when he saw a deer running leisurely along a barren, rocky slope parallel to the river. Thirty yards behind it was a wolf, moving slowly, keeping its distance, seeming to drift along like a shadow. Suddenly the wolf sprang forward, bounded a few times, and grabbed the deer by the nose. The deer somersaulted, hit the ledge, and broke its neck.

"Instantly," Olson later wrote in *Open Horizons*, "the wolf was upon it and the struggle was over. Only then did it see the canoe; it stood motionless for an instant, threw up its head, and bounded back into the spruces."

He and friends did what every Indian and voyageur would have done before them: They turned their canoe to shore, jumped in the shallows, ran up to the dead deer, bled it, and cut off a haunch of venison for supper. The rest of the deer went to the wolf that was circling warily in the brush.

Olson had witnessed the exceptional wolf sighting, the encounter par excellence—one which Chris and I were spared. For it is in the kill—that which most offends the human senses, that which most grates against the wolf's romantic image—that the wolf survives. Here perhaps is the necessary climax to what Barry Lopez in *Of Wolves and Men* calls the "conversation of death" between a wolf and its prey: the possible recognition by both at the start of a fatal encounter that the circle which is integral to all life is inescapable.

	The climax, unlike the kill that Olson saw, is rarely clean. There is blood and bawling and viscera. Killing-to-live cannot by its nature be dainty. But it is the way things are.

I thought of all this as I followed the fresh wolf tracks near my cabin at sunset, tracks which led me down to the lake's edge. My heart, in one sense, was heavy. What would Bojo—had she lived—look like? Would she have come when called? Would she have been lead dog of a sled team? Would she have sung with me, howling, beneath stars? Would she have been this man's proverbial best friend?

	Painful questions, these. Impossible questions.

	Certainly I was justified in dropping them as I headed out onto the lake. There, more wolf tracks meandered—almost undulated—from side to side through a narrows at the east end of an island. Soon they grew faint beneath windblown snow, and I was faced with an alternate vision.

	Bojo had known the freedoms and joys, the open spaces and wild scents, of wolf country. Here she had known the pleasure of rolling on her back among white blueberry blossoms, of snapping playfully at butterflies, and ogling eagles from the cabin yard as they circled overhead. Here she had never known a cage or fence. And here—as she lived and died on wolf country's terms—she had inadvertently led me closer to the mystery and power of *Canis lupus*.

	It was against this backdrop, this land still alive with wolves, that the beauty of both Bojo and her alleged assassin so readily stood out.

	I turned, finally, toward home.

FOSTER FAMILIES FOUND FOR DESERTED CUBS

Lynn L. Rogers

The nine-pound orphan cub clung tightly to my neck as we sped along scenic snowy trails in the Superior National Forest. He looked ahead into the warming April wind, seeming to enjoy the only snowmobile ride he would ever have. He and his sister had been abandoned a month earlier in Michigan when hikers discovered the den where the cubs were beginning their third month of life under the warmth of their mother.

The hikers and other people had returned day after day to see, photograph, and videotape the wild family. Once people learned the mother was more docile than the mother bears of books and stories, they became bolder. They made noises to get the mother's attention. They prodded her to see the nursing cubs. Finally, their harassment became too much for the mother. She abandoned the den.

A day later a wildlife official tracked the mother for more

than a mile through the early March snow. Her trail continued away with no sign that she would return. The official rescued the cold, hungry cubs. In an effort to save the cubs and return them to the wild, wildlife officials began a widely publicized search for a foster mother.

Black bears that are still in their winter dens will accept strange cubs and raise them as their own, teaching them locations of wild food patches and giving them all the advantages of the mothers' territories. However, after a month of searching, no suitable mother was found in Michigan. Wildlife officials became concerned. The cubs had been away from their mother long enough that they might no longer accept a wild mother, even though human contact was being kept to a minimum.

Finding a wild mother was the only feasible plan because the officials had learned by then that no zoo in the Upper Midwest was prepared to raise the cubs. Time was growing short. Bears were beginning to leave their dens, and mothers outside their dens would soon become more discriminating and might reject or kill strange cubs.

Rejected Cub

Wisconsin Department of Natural Resources bear biologist Bruce Kohn found a wild mother that had recently left her den. With no other choices available, Michigan officials flew the female cub to Wisconsin where Kohn released it near the mother. The mother heard the cub, investigated it, but would have nothing to do with it. Kohn retrieved the cub and cared for it while the search was broadened to northeastern Minnesota, where bears emerge from their dens a couple weeks later. Fattened by 1988's abundant crop of hazelnuts, these Minnesota bears were unusually fit for adopting orphans.

A call from Michigan DNR regional wildlife manager Gary Boushelle to Blair Joselyn, Minnesota DNR wildlife population and research manager, cleared the way for the possible adoptions. A day later, the cubs were flown to Ely, where wild mothers were being studied as part of the U.S. Forest Service's North Central Forest Experiment Station wildlife habitat project.

The male orphan and I snowmobiled to the den of a 175-pound seventeen-year-old mother that had two cubs of her own. I stopped a hundred yards from the den. The cub loosened his grip on my neck and stepped up on my shoulder, standing with his front paws on my head, to look around.

I set him on the hard-crusted snow and did a sweep with the telemetry antenna to see if the radio-collared mother was still at her den. She was.

The cub followed close behind as I walked to the den. Soon, I saw the den, a surface nest similar to the one the cub had been born in. This nest was in a thickly wooded spruce lowland, next to the dark upturned roots of a windfall. The mother was sitting upright in the nest, sideways to me, watching with an uninterested, lethargic look.

I adjusted my path to avoid frightening the mother bear with a direct approach. Ten yards away, I picked up the cub and gently tossed him halfway to the den. He yelped as he plopped on the snow, and the mother suddenly leaped from the nest, bounded over the windfall, and tried to gather the cub to her, grunting with concern. The terrified cub screamed, turned on his back, and fought her with all four feet. She turned away, and the cab scampered back to me. The mother, scared by my presence, moved off. I put the cub in the nest with the mother's two cubs, but he was afraid of them, too, making the threatening gurgling sounds that adults use when a fight is imminent.

Defensive Huffing

I left but watched through the trees as the cub left the nest and climbed a tree. The mother returned, checked her cubs, and then climbed the tree, again grunting her concern for the cub. The cub huffed and blew and chomped his jaws in fearful, defensive threats. The mother then took her two cubs and led them away.

The next day I returned to find the cub up a different tree, the mother's tracks under the tree, and the mother and her cubs at a tree seventy-five feet away, patiently waiting for the new cub to join them. When the cub saw me, he descended the tree and tried to climb my pant leg.

The cub clearly preferred people to bears, behavior that suggests the third month of life is important to the development of social ties. It is then that most bears leave their dens, and the attachment between mother and cubs becomes all important to the safety of the cubs. Cubs normally view animals outside their family as dangerous. This cub, having lived with people his third month, apparently viewed animals other than people as dangerous.

I pushed the reluctant cub back to the mother. She ran to him. He squalled and ran back to me. I gently tossed him past the mother. He screamed, unable to get to me without going past the solicitous mother. I hurried away before he could get around her.

When I returned the next day, the mother was standing guard under a tree with *three* cubs in it. The tired and hungry cub had finally given in and accepted the mother's offers of warmth, food, and protection.

The orphaned female displayed a similar fear of bears when I tried to present her to a wild four-year-old mother. This mother had earlier learned to accept human presence and was the subject of intense ecological research by biologists at the North Central Forest Experiment Station.

After being pushed toward the mother bear, the cub ran back and climbed the pant leg of Ugo, a visiting Italian who spoke very little English. His look at me was priceless as the mother carefully peeled the clinging cub off his pant leg with her mouth. The cub soon gave up and accepted the strange mother.

Fate of Orphans

Over the next months and years, we will learn whether it is feasible to return orphaned cubs to wild mothers after the cubs develop an attraction to people. To learn the fate of these two cubs and their families, we plan to visit each family's den next winter. Until then, we will leave the male cub and his new family completely alone but periodically visit the female cub as we continue our study of her mother.

So far, the female cub still likes people. When I visit, she whines to climb my leg and relax on my shoulder or lap. But after a few minutes, she goes back to sniffing and tasting her surroundings, napping, nursing, and watching her new mother forage. However, several times when I've slipped away to leave, the cub has discovered my absence and homed in on the rustling of my footsteps. More than two hundred yards away, I'll suddenly find the cub pattering along in hot pursuit with the anxious mother close behind, calling her back. When the cub stops near me, the mother either picks up the cub or grunts for it to follow, which it eventually does if I stand perfectly still long enough. Sometimes I hand the cub to her and she gently closes her mouth around its head or shoulders and carries it off.

Both foster families live deep in the forest where people seldom go. When a person or other form of danger is heard, both mothers tree their cubs, stand near the tree, and try to identify the danger. Nevertheless, a possibility remains that the friendly cubs might hear people and run to greet them. Not understanding the behavior of the gentle mother who follows, grunting to her cubs, a frightened person might shoot the mother. Black bear mothers rarely defend their cubs against people. The grizzly bear's reputation for defense has carried over to black bears, and black bears at times reinforce that notion with ferocious bluffs.

If the female cub and her mother survive and continue to tolerate people, they can show researchers more about black bear life than could ever be learned otherwise. The detailed information these bears can give now and in years to come can help forest managers maintain proper habitat for black bears and give people a better understanding of all aspects of black bear life.

THE WOLF AT THE WINDOW

Ellen Hawkins

Thump . . . thump . . . thump!

Loud and insistent knocking shocked us out of our sleep. Who would come to our snowbound north-woods cabin in the middle of the night? Gary grabbed the flashlight and hurried into the kitchen. Then, realizing that the sound was not coming from the porch entry, he swung the beam toward the living room's ground-level windows. I caught up just as he froze in his tracks.

"My God, Ellen, it's the wolf!"

Stunned, we stared at the face pressed against the glass, at the blazing yellow eyes and broad cheek ruffs of an adult timber wolf.

This wolf was not a stranger to us. I had first seen him a week before, curled up by a deer carcass in a clearing at the base of our ridge.

During the coldest winter months the Minnesota Department of Natural Resources supplies us with road-killed deer for use as wildlife food. We pull the deer remains by toboggan to a clearing two hundred yards below and in full view of our house. It's a place where wild animals can feed and feel secure—a mile from the road, with Superior National Forest all around.

Blue jays, gray jays, ravens, foxes, fishers, martens, and weasels feed on the deer carcasses throughout the winter. In spring, to our delight, they are joined by bald eagles and turkey vultures. We find wildlife watching endlessly entertaining and a great education.

Wolves stop here rarely, and we feel lucky to hear them howl or to find their tracks. Although this part of the state is their only stronghold in the nation, outside of Alaska, they are not common even here. Their numbers are stretched thin across thousands of acres of forest. This scarcity and their shy nature mean that a glimpse of a timber wolf is a rare occurrence and that a chance to watch one is a special treat.

When I first spotted the wolf at the deer carcass I excitedly watched from the living room window and wrote in my journal as he awoke from his nap, licked his feet, and stood and stretched.

December 8: "The wolf is finally up, and I can see that he's quite dark, his guard hairs black-tipped gray, with lighter eyebrow spots and cheeks, and reddish fur behind the ears. His tawny legs seem spindly above those great big feet. And he has a radio collar! I didn't know there was anybody studying wolves in this part of the forest.

"A wolf does something magic to the place where he is. Here is the same familiar scene, the dark edge of the forest meeting the bright snow of the clearing, the big spruce in the foreground and the vertical lines of the young aspen thicket to the east. But now the wolf is here and there is a vital focus. The frozen scene is charged with life."

But my great excitement at having him here soon became subdued. I wrote:

"This is a hurt wolf. He holds up his right front foot and limps. A couple of times he has fallen in soft snow on raven runs between the old deer carcass at the edge of the woods and the fresh carcass in the clearing. His tail is tightly down, except when he's after the ravens, and then it's held out only slightly.

"His movements seem stiff and awkward. He's very thin. Standing, facing away from me, his body looks narrower than his head. And he doesn't seem enthused about things. He is droopy, indecisive, unhappy.

"When he first woke up he did some grooming, but otherwise he's been lying down, either tightly curled or watching those pesky ravens or looking around in a desultory way, ears drooped slightly back.

"The ravens are getting braver, and he can hardly stand seeing them at a deer. They come dropping down out of the trees around whichever deer he's not at, and he has to hurry over to get them up. They scatter briefly, but here they come again, settling down around the other deer, and back he has to go."

December 10: "The wolf is still at the deer. In fact, if he ever leaves it, it must be at night. He seems weaker and has long since conceded the older carcass to the ravens. Now his only means of defending the other is to lie on it. Even so, they come sidling closer, stand nonchalantly around for a while, then step up closer yet. The wolf curls his lips as he watches with his head on his paws. Suddenly the ravens scatter in a mad flapping of black wings. The wolf must have snarled. But a minute later the birds are back.

"I watched as Gary went crunching down the snowshoe trail, out of sight but not out of hearing of the clearing. As he reached the bottom of the hill, the wolf stood up and watched alertly, ears focused, but a moment later he was relaxed and lying down again."

December 12: "This morning, the wolf's fifth day in the clearing, it seemed time to take him another deer. He didn't seem to have as much energy as he's had; there wasn't much vigor in the way he would shake and stretch. He had been spending most of his time lying down, and at times appeared to be coughing. Even if the old deer carcass had some meat left, it would surely be getting hard to reach.

"Until now, we'd tried to stay out of sight, wanting him to feel comfortable about staying with this easy food source. Now I stood in front of the house where he could see me, hoping that somehow he would be used to us. He must have had glimpses and smells of us over the past several days. But no. He got right up and hurried into the woods, looking back over his shoulder at me. I skidded the deer down and retreated, but he hasn't returned."

December 13: "The wolf is back. He's eaten a little, but has spent most of the day lying between the two carcasses at the center of the clearing. He makes no effort to fend off the ravens, and they are all over both deer."

Near sunset of that day he was gone. Ten hours later we were confronted by his face pressed against our window.

As we stood gaping, too astonished for the moment to do anything, we saw the wolf's nose once more thump hard against the glass before the face withdrew from the circle of light. We heard crunching steps in the snow at the corner of the house, then silence. Was he gone? We scraped a hole in the frost on the south window and again found ourselves trading stares with the wolf. The roof of our pit-style greenhouse is attached to the house just below the south window. The wolf had climbed a snowdrift onto the greenhouse and now sat leaning against the window and looking back over his shoulder at us.

Now came a flurry of activity: getting together chicken leftovers, gravy, butter, and hot water, slipping on our parkas, and hurrying out to see what the wolf wanted. Gary tossed the chicken onto the greenhouse roof and pushed the gravy pan up to him with a snow shovel. I stood behind Gary with the flashlight, the backup person. We didn't think a normal wolf would attack a person, but this wolf was doing something we had never heard of a wolf doing. We didn't know what to expect. The wolf just watched, looking alertly first at us, then at the food.

Now another flurry, this time of indecision. Was he hypothermic? Did he want to come in? Should he? How could we get him in, anyway? We certainly couldn't just leave him there, this wolf we had watched and been concerned about all week. On that still, moonless night, the temperature was twenty-five degrees below zero. Surely it would help to get him into a warmer place. Gary got a blanket, went along the edge of the greenhouse behind the wolf, and threw it across the animal's back. The wolf jumped, then settled down. It looked as if it might be possible to catch him. So I went up to the shed to get the stove going, thinking that the shed might be the place for him, and Gary got the old green quilt.

By the time I got back to the house, Gary was coming around the corner carrying a blanket-draped bundle. He had thrown the quilt over the wolf and, getting no adverse reaction, had tucked it around the animal and pulled him across the slippery roof to the edge. He had looked under to see where the wolf was, covered him again, and scooped him up into his arms. The shed was forgotten. I opened the door, and Gary carried him inside. Beginning to lose his grip, he just made it to the living room and eased the wolf to the floor. He lifted the blanket and stepped back. The wolf looked around in a dazed kind of way. Twenty-five minutes after the knocks on the window he was inside.

Now what? First, get Tom. He is our good friend and neighbor, our only neighbor within twelve miles, and we knew he would want to be in on this. I set out on snowshoes for the half-mile trek, welcoming the chance to try to absorb the events of the night. In the starlight the trail was only faintly visible. The cold that tore at my lungs was making the trees pop, the only sound that broke the silence. A meteor shot toward the horizon, where the dark form of Tom's cabin loomed.

Tom was a bit startled by my greeting: "Hurry! There's a wolf in our house!" and was ready to go before I had a chance to tell him the whole story.

Meanwhile, Gary was getting things ready in case the wolf got more active. He put more wood in the stove, partitioned off the living room as best he could, and put breakable things aside. By the time Tom and I arrived, the thin chunks of ice that had covered the wolf's fur had melted off. We watched him sit up, look around, and walk over to the small space between the lounge and the stove. He lay down there, his head and shoulders leaning against the lounge, facing us.

The three of us sat in the kitchen, whispering, wanting to bother him as little as possible. We felt excited, awestruck, and I, at least, a little apprehensive. What was the best thing to do? It was 3:30 on a Saturday morning. We thought our chances of finding a vet or a wolf researcher who could come would be remote. The attempt to find someone would mean a half-hour snowshoe to the truck and, if the truck started, another half-hour of driving to the nearest phone. We decided to wait until daylight.

Our peeks into the living room were returned by a steady gaze from those bright golden eyes. Pretty soon the presence of a wolf in the living room was irresistible, and we went in to sit on the window bench just to be closer to him.

Quietly, we kept company with the animal that had always seemed to us to represent the essence of wilderness. And we puzzled over the events that might have brought this creature to our home. His radio collar suggested an answer. Maybe he had a history of contact with people that went beyond a single encounter with a wildlife research biologist.

He certainly gave no indication of being upset by our presence. Twice Gary stood beside him to put more wood in the stove, and the wolf continued to lean against the lounge. We know wolves are social animals, able to communicate with facial expressions and posture and vocalizations. But this wolf merely looked about and took a few interested sniffs at pans of water and meat scraps. Still, he was alert to strange human noises: the clattering of pans, the fire burning, our talking. We began to feel hopeful. We had wonderful fantasies about a shy but friendly wolf recuperating with us until ready to return to the wild. We spent long moments just admiring him: the impressive breadth of his handsome head, the lush bunches of face ruff framing those compelling eyes, the grizzled fur luxuriously thick, right down to the black tip of his tail. Until now, our main contact with wolves had been restricted to seeing their tracks along our trails or following the frozen river, so we were especially interested in seeing wolf feet close up. Between long, supple-looking toes grew feathery tufts of reddish fur. And now we could see that he had lost part of a front foot. Had he been unable to hunt because of this? Maybe this was his problem.

But we gradually became aware that his breathing, a little wheezy from the first, was becoming worse. After an hour or so, it was a terrible, deep gurgling.

Gary went over and sat on the lounge. Gradually he moved closer until his hand was right beside the wolf's head. Then he stroked his head and ears. There was no reaction that we could see. Gary put a finger under the collar and thought it too tight for an animal having a hard time breathing. So when the wolf finally stood, with some effort, and slumped down to lie flat beside the stove, Gary used the pliers to take off the collar. Throughout this and all the other strange things that happened to him, that wolf never growled or so much as curled a lip at us, as he had so often done to the ravens. Nor did he act afraid.

His well-worn radio collar was inscribed with the number 6530 and an address of the U.S. Fish and Wildlife Service. Thanks to the collar, we were to find that this animal truly was a wild wolf. But more than that, we were to have a fascinating glimpse of his roots.

For more than twenty years research biologist L. David Mech has been studying timber wolves, first in Lake Superior's Isle Royale National Park, and more recently in northern Minnesota. One technique that he and his associates have perfected for wolf studies is radio telemetry. Wolves are trapped, anesthetized, radio-collared, and released. They can be located and observed by people with monitoring equipment. This yields invaluable insights into all aspects of wolf ecology, information that would be difficult or impossible to gain in any other way.

Look back in Mech's records, to December 1973. A female wolf moves through the forest east of northern Minnesota's Iron Range. She has traveled alone for more than a year, roaming across vast stretches of the forested, rocky lake country of Superior National Forest. Her route has encompassed some twenty-five hundred square miles.

An encounter with a wildlife biologist's trap has left her with a radio collar that reveals her locations to aerial researchers and with a number for their records: 2473.

Now she meets a lone male wolf, and together they set up a territory. It meets their needs. Undefended by other wolf packs, its forty square miles provide adequate food for them and their pups. And in this sparsely settled area, activities of people and wolves will seldom conflict. In the spring, she bears their first litter of pups, and they become known to researchers as the Perch Lake Pack.

Four years later the Perch Lake Pack is thriving. Wolf 2473 has a daughter who has become the pack's new alpha, or dominant, female. Her mate is the male who moved in after her father disappeared during the hunting and trapping season of

1974. From now until 1985, these two will be leaders of the pack and parents of all the offspring, many of whom will be carefully studied by researchers.

Three pups born in the spring of 1982 were radio-collared. As most young wolves do, they eventually left their home territory, at different times and along varying routes. Male Wolf 6441 left the territory in May 1983, when just over a year old. Eight months later he was killed by a trapper in Ontario, 115 miles to the northeast. Female Wolf 6443 left when she was a year and a half old and settled just southeast of her home territory. She found a mate, but after their effort to raise pups apparently failed, she returned to her home territory alone. Since then she has been in the territory just to the south and west, usually alone.

Their brother, Wolf 6530, stayed with the pack until he was nearly two years old, and then he went traveling. For three months he investigated the forest just to the west. Then he headed northeast, and by August he was near Alice Lake, about forty mile from home. He stayed in this area for five months. Researchers hoped that he would find a mate here and establish his own territory. But early in 1985 he returned to the Perch Lake Pack; he remained with them for two months, passing the date of his third birthday. By June he was back at his Alice Lake hunting grounds.

That same month, on a canoe trip in the Boundary Waters Canoe Area, we were thrilled to find fresh wolf tracks and droppings on a portage that we later learned was just three miles from the point mapped by researchers for Wolf 6530 at that time. Much as we would like to think that our paths crossed then, we had seen wolf signs in that area before. It is possible that while there Wolf 6530 found himself unwelcome in territory already occupied by a pack, and that that is why he resumed his travels. Aerial tracking spotted him twenty miles to the southwest in July, fifteen miles farther in August, and twelve more miles to the southwest and very close to his home territory by September.

But later that month, Wolf 6530 covered the distance back to the Alice Lake area, and he was still there in October. Then his signal was lost. His whereabouts were unknown until he found the frozen deer carcasses at our place, twenty-five miles southeast of his last recorded location and forty-five miles east of the Perch Lake Pack.

Now young, well-traveled Wolf 6530 lay on our floor, frightfully sick. His breath came in growling wheezes. Again he got up, painfully, and leaned right against the stove. The smell of singed hair filled the room as Gary rushed to pull him away. The wolf staggered to the center of the room and collapsed. Three times he stretched across the floor in rigid spasm. After each horrible rattling breath came a terrifying moment of no breath. Then, convulsively, a gurgling gasp. But he hadn't the strength to cough and clear his lungs. He wasn't getting enough oxygen. His lips and tongue turned blue. We sat close beside him and strained with him as he struggled to breathe.

He pawed at his mouth. He tried to get up, and did, partway. He lay with his head up and breathed more easily, but then came another convulsive wave. We hung on each long moment between the breathing out and gasping in. But then one of the moments stretched out way too long, and Gary's touch could no longer stir him. Kneeling by his head we could see his eyes change. The focus was lost, and the yellow faded as the pupils became huge. We could look way into them and see the sunrise light reflected in the green blaze.

We sat there a long time, grieving. And we looked at him closely. A thick winter coat had hidden the extent of his emaciation. Beneath it his bones protruded. He weighed fifty-five pounds but should have weighed at least seventy-five. There was a tear on his lower lip, an old wound. His feet were supple, the pads squeezable and spongy beneath their calloused surfaces, and the fur tufts between his toes were silky. The wolf's right foot had lost three pads. On the left foot one pad was mutilated. These, too, were old wounds, noted in Mech's records since 1983. Mech believes they were probably the result of the wolf's getting caught in a fox trap, pulling the trap loose, and wearing it until the toes sloughed off, a hazard for all Minnesota wolves during the trapping season.

We went out to see what the wolf's tracks could tell us about his last night. We found that although he had wandered all over the woods behind the clearing during the previous week and had made seven beds there, this last night was the only time that he had come toward the house. He had walked to the base of the toboggan run and had climbed all the way up the long, steep hill and then followed the snowshoe trail around behind the building.

There he stood, then turned and went back down to the clearing. Later he had curled up under the big spruce tree just long enough to make a slight depression in the snow. Then he came up the hill again, all the way up that steep killer of a hill, taking small steps but no sitting stops until he was again behind the house. There he shook himself, and bits of lichen and twigs that had clung to his fur flew across the snow. Then he turned onto the terrace, walking close beside the house, and squeezed between the bench and the house, where a tuft of woolly wolf hair still hangs from a bent nail. And then he was at the window with that haunting wolf face and those insistent thumps.

After our outing we were more puzzled than ever about what might have prompted the wolf's visit. He had spent so much energy when there must have been little energy left. If it was a random, delirious act, it seemed more likely for him to have gone in any direction but up that steep hill. We snowshoed out to the road and drove to town to call Dave Mech, hoping for some answers.

Mech knew exactly who we were talking about. The wanderings of Wolf 6530 had been of great interest to him, and this information about the wolf's final travels and death would be a valuable addition to his records. An autopsy later revealed that this wolf had died of a fungal pneumonia, and Mech was able to document a natural cause of death that had been virtually unknown among wild wolves. To have contracted the pneumonia, Wolf 6530 must have been stressed in some way. Quite possibly he was undernourished to begin with. Prey animals—few and far between here—are hard to come by in the winter, especially for a lone wolf suffering the nagging pain of an old injury.

We were intrigued by the story of the Perch Lake Pack. And it was a comfort to know more about Wolf 6530. We had developed a deep feeling for him from just one short but intense experience. His story gave us more of him to know.

Mech was able to say something about the great mystery of why the wolf came to the house. Incredible as it seems, he said, it is not at all unknown for starving wild animals to come to human habitation: wolves, raccoons, bobcats, and bears have all approached people near the end. Perhaps these animals sensed warmth or food. Yet these explanations don't adequately fit this case. The house is too far from the clearing for warmth to be felt. As for food, there was still plenty of meat left on the deer carcass.

We'll never know what motivated him to come our way. I can only say that I'm grateful to Wolf 6530 for sharing his last, desperate moments of life. His act gave us a sense of connection with his world that we would never have had, and our commitment to live in harmony with that world has been strengthened. We will always carry with us the vivid image of the wolf at the window.

New Sport for an Old Hunter

John Henricksson

Maybe you see a set of deer antlers mounted separately, a carbine length apart, to form a gunrack on the paneled wall, or a single antler screwed to the cedar lintel above the woodshed door. A hunter's cabin, most likely.

But look closer. What does the base of the antler look like? If it has been sawed off, or is still attached to the skull, it is likely a hunter's trophy, but if the whitish, slightly mounded sponge of the antler core is visible and ringed by a coronet of tiny pebbles, you're looking at a casting—the discarded antler of a liberated and grateful deer.

There is an antler like that on my cabin wall and I value it highly, not only for its perfection and beauty, but for the liberation and gratitude it brought me, the hunter.

I am of a generation that spent its childhood years in what has become known as the Great Depression, an era when economic conditions in this country were severe and family budgets were in deep weeds most of the time. A few ducks, a pheasant, or a couple of rabbits were a welcome sight to many mothers who often wondered how to cope with the appetites of a growing family. The annual deer usually came just in time for Thanksgiving in many rural areas. Youngsters became hunters and providers at an early age. Most twelve-year-olds in my part of the country could make a good overhead shot on a speeding teal or snap-shoot a running deer in an alder swamp. It was something most boys learned as a part of growing up, like how to slide under the catcher's tag or how to sneak into the circus. It was great fun, of course, but the concept of hunting for sport came later when game and habitat became much more scarce.

Now, in the Boundary Waters area of northeastern Minnesota where my cabin rests on the shore of Gunflint Lake, we have a comparative abundance of game animals but I tend to view the deer, the moose, and the ruffed grouse more as neighbors than as quarry. This doesn't mean an antihunting attitude; I'm aware of the conservation work done by many hunters each year. It's a personal thing. I just don't care to pull triggers anymore.

I haven't missed the killing, but how I missed the hunting . . . for a while anyway. Those long days in the autumn grouse woods, jump shooting mallards from a johnboat floating down the creek, the rough comradeship of the deer shack. Golden days. However, there were some other things that stuck in the recesses of memory: the slow misting of a shot deer's eye and that moment when a ringneck rooster towers out of standing corn, filling the small sky over the twenty-gauge barrels with splendor. The barrels jump, the bird crumples, and the king is dead. Triumphs once, but I don't see it that way anymore.

A few years ago I discovered an acceptable alternative that has taken me over the game trails of the winter woods with the same sense of wonder and anticipation, the same hunter's eye and sometimes with the same fulfillment. I am a hunter once again, but my trophy now is only the antler of the white-tailed deer shed in the early winter by an irritable buck as he rubs and twists his rack through the pliant young trees.

This sport of antler hunting has some great and often overlooked advantages. The season is practically year-round. There is no equipment cost. There is no carcass to clean and drag a country mile out of the woods, and there are no stray bullets from careless hunters whining overhead. The forest is always a fascinating place for those who worship there, listening to the audible silence of the snow-muffled woods or sharing a sandwich with the brazen Canada Jays. Occasionally, one sights a lordly moose up to his knobby knees in fresh snow, his myopic eyes squinting, the great head swaying from side to side testing vagrant breezes. This is high-quality time, and once in a while I return to the cabin with an antler slung over my shoulder or a spike tucked in my belt.

The buck has grown those branching beams between April and September to impress the does with his masculine charm and sometimes to use as weapons in the sparring matches he provokes with other bucks who may have designs on members of his harem. In the summer the antlers are covered with velvet, a form of skin that brings a blood supply of protein and minerals to the forming bone.

The velvet is shed in long, stringy rags in late summer and the buck spends a lot of time then rubbing and polishing, primping and prancing, anticipating his coming social calendar.

After the rut, or mating season, the antlers begin to demineralize and the buck impatiently rubs and scrapes them against trees to dislodge them. Casting times vary from north to south and according to deer biologists, the shedding of the antlers is determined by the number of daylight hours. Here in the far north the casting starts late in December or early January, whereas in Mississippi, the average casting date is April 7.

The deep winter here in Minnesota is the best time to go antler hunting, although later in the spring when the snow begins to melt, they are often more visible. Summer isn't a good time because of the heavy ground cover, and, by fall, even though the woods are bare of foliage, the mice, shrews, and other rodents have usually recycled them for the calcium and phosphorous they contain.

Quite accidentally I scored last year because of a late March thaw when we had about forty inches of snow left in the woods. At that time of year we often get a couple of bluebird days or a warm rain that will melt enough snow cover to expose antlers discarded along the trails. The warm March wind had swung to the east when I strapped on the snowshoes and took off on a well-worn deer trail that led to a cedar swamp yarding area I wanted to search. A group of deer from this yard had visited our corn pile often during the winter, and on Thanksgiving Day a blocky gray buck with a beautiful eight-point rack showed up right at noon. Naturally, we referred to him from then on as our Thanksgiving buck.

On this March morning I wasn't more than a hundred yards from the cabin when I spotted a polished tine about three inches long sticking out of the melting snowbank on the lee side of a white spruce grove. Like an archaeologist at a dig site, I carefully brushed snow from around each side and eventually brought out a perfect four-point antler without a scratch or nibble anywhere. It was beautifully pebbled at the base and perfectly symmetrical. On the long tine was a little burr indicating the promise of next year's point. I chose to think it was left there for me by the Thanksgiving buck repaying our hospitality.

A buck will often cast both antlers along the same trail, which doubles the chances for the ultimate trophy, a matched set. A few of my fellow antler hunters have found them but so far they have eluded me. However, being a hunter once again, I can always say, "Wait until next year."

CANOE COUNTRY

Florence Page Jaques

Wednesday, August 31st

These voyaging days are translucent with joy. When we start out in the morning, the earth has such a before-Eden look that it seems a shame to shake the dew from the blueberries or strike our paddles into the sleeping water. Thrusting on into sun-filled channels; drifting into green-needled embrasures where chickadees are buoyant; landing on a beach to bathe and to read the overnight paw prints—it is all intoxicating.

Now we have left the smooth pine slopes and the great bare hills of stone. We have come to rugged shores, ancient pines, and huge creviced rocks, rich in tone, padded deep with hoary moss and gray-green lichens. When the lakes are calm, they reflect the most glowing colors, dark wines and purples and crimsons, deep blues and greens, that we haven't noticed on the banks.

Only after we have seen the pure colors in the water do we look up and distinguish them in the tree trunks and cliffs, where we have seen only browns and greens because we weren't expecting anything else. An artist once told me that if you really want to see the color of a landscape you should stand on your head. This of course is practically the same thing.

We are camping on a jagged island in Lac la Croix.

Until now I have been very amenable about our camps. I've liked to go along and never know when Lee would say, "Why not camp here?"

But that is changed. Yesterday on the Curtain Falls portage—by the way, have I said how deeply I approve of portages? After miles of sunny waters, to have a chance to use your legs instead of your arms for propulsion, to plunge into a crisp, shadowed path, sundering ferns and bushes, brushing spiced boughs, a turquoise lake behind, unknown water ahead, feet clinking on the stones! In our doubled trips across, I never get quite used to leaving half of all our worldly goods sitting unprotected on the rocks. Nor can I ever take for granted the wood fragrance, so different from the smell of the air on the lakes. I have pine scent inextricably mixed with portages now; one will always make me remember the other.

As I was saying, on this portage we met a couple who had been fishing for deep lake trout. The four of us sat down by the black curves and foam white of the rapids and talked, mostly about trout. However, Mrs. Morse did mention casually that they had seen no moose this year but that wolves seemed unusually prevalent, and so they had preferred to camp on islands.

So now I prefer to camp on islands too.

I can't help it. Lee teases me, and I know it's foolish, that the wolves can swim if they're really hungry, that they only get ravenous in winter, and that even in winter it is extremely doubtful that they ever attack man. But still Mrs. Morse's words reverberate in the air. And after all, why *not* camp on islands?

They're much more interesting than mainlands, which run on and on in an aimless sort of way. Each island we meet has an individuality of its own, serene or careless or aloof. We saw an innocently wanton one clad in nothing but two daisy plants.

Our present island was bequeathed to us by the Morses.

It's just beyond Bottle Portage. It may be the one where Macdonell "Killed a cub Bear and slept in sight of the Mai." The mai was a lobstick or maypole, a favorite landmark of the voyageur, made by cutting away most of the branches of a tall tree. I like to think what welcome signs they must have been, just as the international boundary markers are hailed by us with joy.

This is an island with an escarpment, rather high and harsh, around three sides. To the west we have a small harbor, a semicircle of flat clean-washed stones. The center has an open space for our tent, the rest is inexorably wooded, and the thick dark moss under the trees is wonderful to walk on barefoot.

How much more sensitive we are to feeling in this primitive environment! In fact, all five senses are much more wide awake. Of course, one would expect seeing and hearing to be more enjoyable here, since there are only pleasant sights and sounds. And taste is proverbially keener out of doors.

But how much more I notice the touch of things! The smooth paddle in my hand, the shock of cold water on my face, the texture of rough sticks and pine cones when I pick up firewood. The slimness of pine needles in my fingers, the springy turf or hard granite when I step.

And the increased keenness in the sense of smell. I wish we had a better word for *smell*—one that didn't suggest a disagreeable sensation! *Scent, fragrance, odor,* they are all too sweet, too flowery. No, *smell* is what I mean, the different way that early morning smells, and twilight. The changes one forest path can give you as you walk along, the scent of wet green leaves, and dried brown ones, of mushrooms, and crushed grass.

Last night it was cold. There was a hint of a new moon in the apple-green sky, and a wind in the pines all night long. Snug in my blankets, I could imagine that the branches above us still held faint echoes of war drums, or the songs of the voyageurs:

> Quand j'étais chez mon Père,
> Petite Janeton,
> Il m'envoyait à la fontaine
> Pour pêcher du poisson.
> La Violette Dandon, oh!
> La Violette dandé . . .

But it's time to get breakfast. Lee has gone over to the other side of the island to see whether a certain eagle is golden or only bald, and I've been scribbling, far too long.

Lee will expect flapjacks when he gets back. I'm an expert flapjack maker by now, for our bread is all gone. Lee is better at the actual flap, however, for when I do it, the flapjack, instead of turning the customary mild somersault, soars so high that it is quite cold by the time it comes down.

I cook fish very well now; my boiled beans with bacon and my outdoor coffee are delicious. The only things I cannot cook are the dried peas. I have boiled some for three days, carrying them hopefully along with us, setting them on a fire the minute we stop, and still they rattle stubbornly when the water bubbles. I mean to give them to this chipmunk who thinks he helps me cook. He is the smallest one I have ever seen, and the most engaging. And I'm afraid he knows very well that he is, for he poses outrageously. Now he is sitting on the very tip of a dead cedar bough, eating a prune seed, and only glancing at me fleetingly over his minute shoulder. Sublimely preoccupied!

Thursday, September 1st

We were marooned yesterday on a sand beach. We went to see some more painted rocks, even more interesting than the first cliff, with moose and men in war canoes and the prints of many hands, all in bright red.

These paintings make the primitive red man seem startlingly alive, much more so than seeing modern Indian culture ever has. Especially these prints of hands. Real hands . . . I had a ghostly feeling.

While Lee was trying to photograph the paintings, the wind had streamed down against us harder and harder. White plumes filled the lake till Lee thought it was hazardous to attempt the passage back to the island.

So we scurried on, around a more or less sheltered shore, and found a short sand beach between two vertical cliffs. It was a mere dune between the lake and a swamp beyond, but we took refuge there.

Climbing up ponderable rocks of gray and green and orange to a small cave garlanded with vines, we ate a lunch of cheese and chocolate, which Lee happened to have, luckily, in his pocket. Below us the waves pounded against the sand.

Now, as we sat in our cave, sheltered from the strong gusts, a whole family of moose birds gathered around us, looking at us with friendly eyes, cocking their black-capped heads inquisitively.

This fluffy gray bird is a jay, the Canada jay, also called the camp robber and Whisky Jack, the last derived from his Indian name Wis-ka-tjon. He seems to have no fear of man at all, though he chooses to live in the wild woods instead of around settlements. I had been wanting to see a moose bird, having heard so many tales of his brazenness and robberies.

But I decided today that he is misjudged.

Crows are crafty and clever thieves, taking an ironic pride in their knavery. But the moose bird is so wide-eyed and innocent—I can't believe he thinks he is a robber at all; he just borrows things, feeling sure you won't mind, since he is a friend of yours.

These particular birds sat near us companionably. After a little, Lee set out a crumb of cheese and a crumb of chocolate on the rim of the ledge. Both were snatched up at once. Before long the family was diving and dipping past us, catching slivers of chocolate or cheese or silver paper from our fingers, and perching along the vines. We enjoyed this entertaining in a cave!

After our refreshments had vanished, our guests did too, and we climbed down and explored the hill east of the swamp, till a red squirrel, singing a real little melody up in a tree, saw us and began to scold instead. Really, the animals are as stern with us as they were with Alice in Wonderland. We ruefully made our way back to the beach.

I sat down on the sand and thought delightedly of being reprimanded by a squirrel. I began to realize that one of the deepest joys of this vacation, which I had scarcely been conscious of, is the nature of our social contacts. So clear, so direct—a squirrel's dislike, the wordless friendship with the jays, the comedy offered by two ants with a flag of dried beef, the awe an ancient pine awakes. After the complicated stresses and emotions a metropolitan day engenders, gatherings where intricate attractions and repulsions web the air, confidences given, advices asked for unsolvable problems, faces in the subway with expressions that tear one's heart. Of course one wouldn't want to escape the complex demands forever. But for a breathing space—what a release!

While Lee began a color sketch of the cliff, I lay in the half-shade of tall grass, by a twisted ash tree. When I pillowed my head on my arm I could look along the honey-colored pebbles into a medley of wild-rose briars. Swept by the strong south wind, lulled by the assault of the waves against the sand, I remembered drowsily the first time I ever discovered that poetry had the power to give me cold tingles up and down my back. I could hear the golden voice of our tall young English teacher as clearly as I had when I was thirteen:

> And answer made the bold Sir Bedivere:
> I heard the ripple washing in the reeds,
> And the wild water lapping on the crag. . . .
>
> I heard the water lapping on the crag,
> And the long ripple washing in the reeds.

With that murmur in my mind I drifted off to sleep.

When I woke, the wind was more boisterous than ever, and the waves were storming in, so I went wading down the shore. Of course I got drenched, but it didn't matter; I simply went in swimming. It was glorious! The lake glittered darkly blue, the pines were emerald sequins flashing in the wind. Long tangles of water grass wavered in gold-brown streamers about me, and then the brilliant sand shone clear again through the surge. The waves, rough as half-grown puppies, played about me, tossing me over, pushing me back to land.

The forests were more patrician than usual this morning. Narrow gold diagonals fell through the green-towered ramparts, and the air was cool and touched with fragrance. The water rippled clearly, darkly; outside a shadowy cove the sky to the west bloomed with pale clouds of lavender and faint purple and that creamy fire opals have. I needed some ritual to follow, an orison of worship.

Today we saw our first bear! Lee made me jump by laughing suddenly; he saw it slip from the bluff it was climbing. It fell into the water with a splash, surprising itself and Lee enormously. I was only in time to see a dark shape scramble up the bank, and later, by paddling hard, we had a glimpse of its grotesque silhouette against the skyline for a moment. Lee said that bears will not condescend to hurry if they know you are watching them, but they make up for it as soon as they think they're out of sight.

Later in the morning a gale came up again while we were crossing Iron Lake. The water turned a deep violet, the waves were urgent and white capped. We had to struggle violently to make headway; it was fun, but it was a real combat. Cloud shadows and flashes of sun whirled by us, as we dug our paddles furiously into the surging assault.

"I'll quarter against the wind," Lee shouted to me. "If I drive straight into it, the waves are too close together. We hit too hard."

I was completely exhausted by the time we finally rounded a rugged point and came to smooth, washed shores of pale ivory that were banked by sooty jack pines in close thickets. Here we discovered the mouth of Bear Trap River.

This was a hidden river of abrupt turns and many boulders; remote and casual, it may not have been traversed for many years. A weight of sun lay on its low shores; reeds and rushes and yellow pond lilies, black ducks, blue herons, and bluer kingfishers bedecked it lavishly.

We had lost the wind. It was midsummer and midnoon. We made our way slowly up the river, until we came to a portage. Here was simply an immense face of cliff, seemingly unscalable, blocking the river, which disappeared in an offhand manner.

Even if it hadn't been remorselessly hot, I was sleepy from fighting the waves all morning. I assembled a modest lunch, though untying the strings of the bag seemed an impossible task. "Hurray," I said to myself, as I gazed at the cliff, "that's stopped us. I can have a nap."

But Lee, fascinated beyond words by the disappearance of the river, was enthusiastic about climbing over the cliff to see what we'd find.

I loathed the idea! I felt fuzzy in my knees, my face was a scarlet flare, and I knew I could never drag myself over the half-mile portage the map promised.

Before I could summon energy to express my violent disapproval, Lee was scrambling up the sheer rock with the canoe balanced on his head; so rebelliously I scrambled too.

We reached the top breathless, and I was allowed to lead the way along a lost path in dense forest. Outwardly a guide, but inwardly a whole mutiny. I hated this path as vehemently as I had loved all our other portages.

Nobody had come here for a long time, we could tell. But there were moose and bear tracks, and I had to scrabble through low branches and tangled places, where I couldn't see what I was going to meet.

I kept stumbling and stubbing my toes. I don't know why I always stub my toes when I am cross; it's one of the minor mysteries of life.

We crossed the portage at last and met the river again. It went placidly on, with grass bogs on either side—a perfect place for moose, Lee was sure. The sun was too hot for me even to mention what I knew perfectly well: there wouldn't be any moose.

There weren't any moose. Nothing alive appeared on that river, except a very drowsy turtle on a rock.

We toiled on down the river, until we came to a place where beavers had been gnawing down trees with industrious teeth and building a supersolid dam. Even to contemplate the enormous amount of labor they had lavished on it made me feel shattered. Infuriating, haggard, short-winded beasts beavers must be, I thought, working like that. They'd be proud of themselves, too, for being all worn out. I disliked them heartily. But at least their dam stopped our expedition.

When we had dragged ourselves back to the luggage and embarked again, Lee was not long in finding a place to camp. He indicated a rocky point.

"Splendid to watch for moose here."

But I felt supremely indifferent to moose, and it was a very small point. The next possible place, a jack pine wood, I refused too; it was too closely overgrown. The third discovery was a bare rock slope with a little thicket, and black forest beyond.

"There!" said Lee triumphantly.

"It isn't an island," I murmured.

"Of course it is. See that marsh—it probably leads around to the river again." He looked so hot, poor lamb, that I couldn't insist on more paddling. I accepted the place as a possible island, and we landed here.

Lee made me lie down in the shade of a scraggy pine, while he made camp. I was almost asleep when he came gaily along, to show me that a bear had been overturning rocks in search of ants, just by our tent. Like Queen Victoria, I was *not* amused. What if it comes back for an ant it has overlooked?

This really is a magnificent view, with the river curving widely on one side, and a great grass marsh on the other. But it's aloof and alien to me; this isn't an intimate place. I'm not a part of it as I've been in our other camps. I don't want to stay here longer than one night.

BOGTROTTER

Richard Coffey

The bright setting sun shined on the towering cloud of a large re-
treating thunderstorm, exaggerating its somber tone; the birch
trees stood tall and hard and white against the blue-black backdrop of rain. Golden leaves fell heavy with moisture and empty of
life. The storm had been noisy, with much cool air on its heels, a
welcome breeze that stirred the forest vapors and gave us our first
rich smell of fall.

It was September when I had first seen the bog. I was
flying a small plane north from the Twin Cities to an early morning meeting in Duluth. Just at sunrise, I happened to look down
to see the first warm rays of sunshine across the bogland below.
The texture of the sphagnum mat was soft and fine like a carpet,
but broken by many small islands and groves of black spruce.
There was little evidence of settlement; the high ground appeared
dense in aspen and birch. I circled the spot on my map.

Late that afternoon, on my return trip, I found the bog

again and, reducing the power of the single-engine plane, I dropped down low over the ancient lake to have a good look. The islands drifted slowly beneath my wings; the air was calm, and the roar of my engine seemed muted, far away. I turned the plane and followed the contour of the woodland shore. Birch trees shimmered pale white in the fading sunlight. I looked overboard from side to side and saw nothing that suggested life. There was mystery below me in the deep shadows that rushed under the plane from all sides.

There was loneliness here, and I felt a chill as another island loomed before me, out of the mosses, into my view. I turned to the south, dipping a wing over the blazing red cranberries, and then I saw the deer. There were eight of them in single file along a narrow trail that connected one island to another. They were frozen, watching the plane, as I pulled away. I longed to walk in that place, and I advanced the throttle. The plane climbed into the sunset, and I marked my map again.

A month later I sat on the front seat of the realtor's car as we bumped along the rough gravel road. We passed small farmsteads carved from brush and rock and surrounded by marsh. We passed hunting cabins set deep in birch woodlots; red shacks trimmed with white and swallowed by the bracken fern. Often we passed a lone stand of white pine; tall and shaggy guards on empty plains. Finally we pulled over on the side of the road and stopped.

"There you be," the realtor said, pointing to a wall of stark white birch trees that stood knee-deep in hazel and tangles of vine and grasses, long dried and brown. The luster of summer had departed the place, but the woodlot gleamed red and yellow in its autumn colors. "There's an old logging road over there if you want to walk in a ways," said the realtor.

What remained of the road was covered with brush, but I could see where the cut had been made by the loggers. Deer had long possessed the trail and it had grown narrow by their use. There was still the faint smell of sweet fern in the cold, moist air, and a ruffed grouse exploded from the dogwood in front of us.

"There's good bird hunting in here," the realtor said, catching his breath. "The oldtimers around here tell me that they used to hunt deer up there on that rise." We pushed through the thickets and gained higher ground. Stands of birch, white in the fading light, rose above us as we climbed the ridge. Fields of bracken fern standing tall as our waists rocked gently in the cool breeze that penetrated the woodlot. A large crow circled above the forest canopy, cawing with a strong voice that echoed on the distant shore. The bogland lay before us.

"There's a good bit of swamp out there, and part of it is on this property," the realtor said grimly. "I just want to be sure that you know that."

"That's perfect," I said. "It's a bog, isn't it?"

"I don't know what you call it, but it's wet and you can't do much with it," he said. "I'll tell you one thing, though: I wouldn't go out there. A fellow wants to be careful about this swampy ground."

We walked to the edge of the bog. A marsh hawk soared low across the mat through the yellow tamarack and into the grove of spruce, a black smudge on the distant shore. At my feet, pools of dark water reflected the pale orange sky above and hid the long, dead grasses beneath the surface. I reached down to wet my hand and smell the rich flavor of this still water. The realtor gave me a suspicious glance, and we turned to walk back to the car.

As we hiked down the ridge we passed bushes of blackberries, the plants of wild strawberry and tall pin cherry trees. Alder was growing in thick stands on the shore of the bog, aspen crowded the trail, and here and there were pools of standing water with cattail and marsh grass. Every few feet the plant life changed as we gained higher ground and then dropped down in low gullies.

"This is wild land, and that's about it," the realtor said. "If that's what you're looking for, I mean."

"It's perfect," I said, ducking under a low-hanging branch covered with lichen. I stopped to inspect the growth, and my companion buttoned his jacket.

"It sure gets cold down here in the afternoon," he said. We pushed on through the brush toward the road.

When we broke out of the woods the sun had set and the full moon would rise in a few minutes. I stood on the road and watched as the golden orb lit the sky behind the spruce and rose above the bog. Then we drove away. Later that night, after I signed the papers for the land, I went home and dreamed. Sitting in my chair overlooking the city lights, watching cars move slowly across the bridge, I dreamed about the lichen hanging from the branches of the dying thorn tree. I dreamed about the aspen leaves quaking in the breeze and the brown grasses that rattled underfoot. I thought of the crow that circled and cawed in a husky voice, and then I slept.

The cloud bank moved eastward, and the sun sank below the horizon. An airplane droned high overhead. Four years had passed since I had seen the bog on that early morning flight. Four years of work in the city, four years of life in the mainstream. We had visited the bog often in those years. We had picked flowers, planted trees and photographed the birds and the deer and the fog at sunrise. We had watched crab spiders change their coloration and ride daisies in the wind. The spiders were waiting for insects as we were waiting for an idea: How could we arrange our lives so that we might come to the woods to live?

From the cockpit I had thought it a dream to live in the woods. Now, standing with my ax on the forest floor, watching someone's small plane fade away to the south, I felt an emptiness that often comes with the realization of a dream. We had worked hard to achieve simplicity, we had given up much to enjoy the peace of the woodlot. The emptiness, the bare spot that remained when our dream became a reality, was filling with a sense that was foreign to us, a sense of the natural world, a sense of the wild. Our pleasures were in the birth of a flower, in the seed scattered by the wind and fed by the rains. Our pleasure was in the egg left by the ovenbird in a trailside nest, an egg that may be food for the skunk or that may hatch and carry the genes another generation more.

Our love for each other grew out of a newfound respect for one another's abilities to contribute to the whole of our life. Our world was stronger and more meaningful because of the presence of each other. At a time when world powers were shouting threats of nuclear buildups, we felt privileged to live in the midst of the rocks and roots and primal ooze on which life depended. It was clear to us that the potential for world destruction was very great. Men and their ideologies were in conflict, and their technological abilities to make war seemed to outstrip their diplomatic prowess. Fifteen billion years ago the Earth may have begun with a bang. What difference would it make to our galaxy if the Earth should flare bright one day and dissolve into dust? What difference if the skunk should eat the ovenbird's egg?

In the light of dusk I filled my wheelbarrow with split birch and worked my way to the main trail back to the cabin. Looking up, I saw, at some distance, Jeanne's white cap bobbing in the gray light. She appeared and disappeared as she moved down the trail. Hidden by the great aspens and diffused as she passed through a hillock of pin cherry, she hiked ever closer, yet the distance between us did not seem to close. Only once in a great while did we realize the true course of our trails. Each trail wandered about the woodlot like a meandering stream, touching the red oak and shooting off to the right where the woodpecker tree invited a turn to the left down to the pond where the spring peepers are likely to begin their spring chorus. We had cut the trails in the summer when the brush was dense, but our enthusiasm was strong. Down on our knees, we clipped away the hazel and the dogwood and inspected the rocks and the mushrooms revealed in our cutting. It was not an efficient way to make trails—armed with clippers and field guide and compass—but we were more interested in what our cuttings would uncover than where the trail would lead. We built the paths to take us to firewood, and firewood is everywhere. So are our paths.

In the failing light I could see Jeanne ahead, dwarfed by the tall birch, kneeling to inspect some small bush. I pushed the loaded wheelbarrow to her side, and she turned from the dogwood plant.

"We have new neighbors," she said.

"Neighbors?"

"Larry and Moe," Jeanne laughed. "Two deer mice have discovered our corn storage in the shed. You have to see it."

We hiked slowly back and stacked the birch on the deck of the cabin. Jeanne led me to the shed where inside a large can that contained corn for the animals were two very fat deer mice. They were chewing contentedly and took no notice of us peering into their lunchroom.

"I found them in here this morning," Jeanne said, "so I put this stick in for them to crawl out. Watch." She picked up a long stick and dropped it into the can. The mice ran up the stick, ran along the edge of the can and jumped to the handle of a nearby shovel. Down the shovel to the floor and through our legs they trotted. One of them went out the door, the other one stopped. He looked up at us for a moment, and then he turned and retraced his path back to the corn. Inside the can again, he picked up a kernel of corn and hastily repeated the evacuation to rejoin his friend.

"I think they've been eating our corn all along," Jeanne said. "But as the level of corn went down, I guess they couldn't jump back out."

"Larry and Moe, eh?" I said, trying to register two more names on our growing list.

The next morning I followed Jeanne to the shed. The deer mice were there, huddled together after a long night of feasting. Jeanne dropped the stick into the can, and the mice climbed out. They looked at us with wide dark eyes and slid down the shovel, homeward bound.

Every morning Jeanne followed the same routine before she filled the birdfeeders, and I could hear her greeting the mice as I built a fire in the cookstove. Finally, the containers needed refilling, and the mice were able to get along on their own until the level was once again too low for them to manage without the stick. Larry and Moe were with us for some time before Jeanne came into the cabin one morning and announced that Moe was missing. He remained on the missing list, and one day Larry was gone, too.

The cabin mouse didn't have a name. She was a white-footed mouse who had arrived in a hurry late in August to build a brooding nest. We met her one evening while we were reading and she was running from shadow to shadow with a mouthful of paper. She ran under the stove and waited. Then she ran to the cupboard and up the wall into a crack. Listening carefully, we could hear her in the ceiling. In a few minutes she would appear again under the stove, then scurry across the rug and into the wastebasket. She worked only after dark, and as she grew accustomed to us she collected her materials closer and closer to our reading chairs.

After we had gone to bed we listened to her travel about the cabin far into the night. When the nest was completed she seemed to calm down and often visited the window near our bed. She would sit for hours on the sill looking out through the screen. As the night air became cooler and we shut the window at bedtime, we always had to check for her and many times coax her back into the room.

One night I looked up from my book to see her at my elbow on an end table. She was perched on a stack of books, staring at me with beady little eyes. Then she pulled a length of dried flowers from a vase and scampered off. We didn't see her for several days, and then one night, when I had stayed up late to do some writing, I happened to see four small mice scaling the wall over the front door. Suddenly, one of the mice lost his grip and fell to the floor with a loud splat. Sure that I was going to have to clean up a dead mouse, I approached the creature with a dustpan. I picked him up and cradled him in my hand. Although he couldn't have been more than a day or two old, he was fat and full-bodied and in every way a white-footed mouse. In a moment he stirred, and I held him close to the wall, which, after a second to clean a foreleg, he scaled with no difficulty. The mice stayed in the cabin for another day, and then they left.

Jeanne and I had never given much thought to the notion that animals and birds could make errors in their judgment. We had always assumed that their behavior was so coded and their highly specialized abilities so expert, that any mistake was certain death. But one morning Jeanne noticed a small sharp-shinned hawk perched uncomfortably on a feeder. He was pinned to the box in fear although there appeared to be nothing wrong with him physically. What was odd was that the blue jays, who usually sounded the alarm when a hawk was anywhere near the woodlot, were quiet. What was even more curious was the great number of jays at the corn pile only a few feet away from the hawk. We watched. The hawk was frozen. Suddenly the hawk pushed off the feeder and climbed up through the canopy of the birch trees. The blue jays went wild. They started a chorus of their hawk call, switched to the intruder call they used when someone came to visit us, and ended the confusion by calling for an assembly. When most of the jays had quieted down they flew off into the trees to roust the hawk. Oddly, the hawk had remained near the cabin and was now faced by a gang of screaming blue jays. The hawk left his high perch and chased a single jay until the others arrived to help. The aerial duel lasted for five minutes, and then the hawk flew off to the west. For the rest of the morning the jays chattered about the adventure. No matter what bird came to the feeder, the hawk call went out and the jabbering began.

We have seen the jays use the hawk call many times to take advantage of the feeders. The call is taken seriously by all birds and animals in our yard. Even the deer are likely to pause in their feeding and watch for the incoming hawk when the alarm is sounded. Squirrels, chipmunks, redpolls, finches, juncos, and chickadees all scatter when the blue jays have spotted a hawk. When the pressure on the feeders is too great for the jays to feast with ease, we have noticed that they will disappear for a moment and suddenly begin the hawk call. When all the squirrels and birds have taken cover, the jays glide in to the feeders. It doesn't seem to matter how often the jays cry hawk, the yard invariably empties when a new call is issued. When the hawks do come into the yard, they are usually successful in their work. We have seen a rough-legged hawk approach the woodlot and move from tree to tree—unseen by the jays—until he is directly over the feeding stations. In a flash he is in the midst of the smaller birds and away with a victim before the jays sound the alarm.

A kill in the yard seldom interrupts the feeding frenzy for more than a few minutes, however. Although the strike is a shock to the serenity of the setting and the jays seem to worry about it for hours, life soon returns to normal and many pounds of sunflower seeds are quickly reduced to dust.

We had lived on the woodlot for six months as September began, and in that time the novelty of the woods and our primitive lifestyle had become routine. We found that the living was easy and peaceful. Jeanne enjoyed cooking on her woodstove far more than she had on the electric model we used in the city. She had produced bountiful dinners and stores of breads and cakes and cookies and pies. Pumping water by hand, walking to the outhouse, heating with wood and reading by kerosene light all seemed natural and right. Living together in a single room, sixteen by twenty-four feet, forced us to practice cooperation, and our relationship was warmer than it had ever been before. We didn't perceive that we were doing without; we understood, instead, that we had lived with excess before we moved to the woods. With little high-value property to guard, to insure or to worry about, we were becoming free. Free to set off on a morning hike to collect berries, free to camp on an island, free to sit in the bush for hours and watch a herd of deer browse in the dogwood. If anything troubled us after our first six months on the place, it was the complexity of the outside world. We feared that an unhappy people with wounded pride would unleash a nuclear war. At a time when we had discovered prosperity in the simple life, it seemed the world was searching for happiness in greater consumption—and was finding instead hate enough to destroy the Earth.

KING WEATHER

Helen Hoover

On a May morning some years ago, I was talking with Awbutch, the daughter of Netowance, both of whom are Chippewas, born and reared in the Canadian forest across the lake. It was a lovely day, soft with the earth-perfumed wind, tender with purple birch buds and new grass. Robins sang, and even the raven that spends the winter croaking dolefully around our cabin was making pleasant coaxing sounds for the enticement of a lady raven, sitting aloof in a nearby tree.

"Mother says snow—big snow," Awbutch mentioned, looking up at the clear blue, where puffy clouds like gold-edged roses drifted. "Did you see the northern lights last night?"

I said that I had, and very beautiful they were—mist-green banners tipped with pink.

"I never heard that lights in the spring meant snow. Did you?" Awbutch asked. When I shook my head she said, "Well— Mother says snow." And we both laughed.

The sun went down in a red sky, the "sailor's delight" kind of sunset. Netowance's snow looked very doubtful.

In the morning the calling of blue jays woke me. I looked out through a curtain of feathery flakes so thick that I could hardly see the trees. The raspberry canes were buried and only the top half of a six-foot elder bush was visible, its new leaves emerging in a surprised way from the snowballs that surrounded them.

The yard was full of birds. Residents that fed with us all winter—chickadees and nuthatches, blue and gray jays, hairy and downy woodpeckers—were calling from the branches, perching on the woodshed roof, peeking in the windows, confident that breakfast would be forthcoming. There were summer birds, too—robins and thrushes, sparrows and warblers and goldfinches—that had followed the lead of the others and were hopping over the drifts in a hopeless search for food. The whiteness of the forest would be dotted with starved and frozen feathered bodies before a thaw cleared the ground. But I could help these few, at least. They could stand the cold if they were fed.

Hastily I broke up graham crackers and crumbled suet, which Ade scattered outside along with seeds and cracked corn. I added canned peas and cherries to the menu and was mixing cornmeal and peanut butter, which robins sometimes will take from feeders, when Awbutch pulled her boat in to the shore and plowed to the house on snowshoes that sank deeply at every step.

"Old bread," she said, handing me a parcel. "To trade for corn. So many birds—they'll die."

When she had pulled her loaded packsack onto her back, we stepped out into the storm. She slipped her feet into the thong hitches of her snowshoes and said, with the infinite contempt of one who has a personal weather prophet, "The *radio* says 'snow flurries'!"

I can never hope to equal Netowance's skill at "feeling" the weather, but I wish I could, because weather is the indisputable ruler of the forest and every living thing in it, humans not excepted. Ade and I gear our lives to its changes, and watch eagerly for the dramas that it brings.

The enormously vital spring ice breakout begins with the first thawing day, when the expanse of snow that glares in the sun from shore to shore loses some of its reflecting power. Gradually it becomes grayish and pools of water lie on the ice, mirroring the sky as pale blue or the cloud cover as dull silver. A fan of silt-brown water spreads from the mouth of the brook. In the distance, for the first time since the previous fall, I hear the faint, skirling cries of the herring gulls, piping in returning summer life.

The leaden ice sheet begins to rot, honeycombing from above and below, until it looks gray-black and sags toward the water under it. Our winter water hole floods and, all over the lake, water seeps up through thin cracks to spread in sheets on the ice surface, freezing at night and thawing in the day.

As the ice deteriorates we look along the shore for cracks. Eventually we see a rivulet of water there, deeper blue under the open sky than the ice-bottomed pools, or more mirrorlike under the clouds, or a richer copper at sunset time. Soon the ice cracks away from the shore and a pair of loons rests on the open water before they fly on, leaving the echo of their calls. More cracks widen, and the network of water gleams violet and indigo against the mat-gray background. As the sun sets, the water turns from gold and copper to lemon color, then to faint green before it whitens and loses the light and is one with the night's dimness.

One day the wind rises. The ice rumbles, snaps, breaks into floes, and begins to move. It may reach shore in huge plates or it may grind itself into fragments first, but its power is inexorable. Poorly anchored docks are lifted and dropped back as piles of planks and logs. Boulders as large as automobiles are maneuvered into new positions. On the windward side of the lake, open water stretches out from shore. Then, with a suddenness that never loses its exhilaration, the last fragments of the winter freeze grind themselves into crystals and tinkle into nothingness. The blue waters advance sparkling in their wake; gulls wheel and cry above the dancing wavelets; a fish flops in the midst of spreading rings. The land may be deep under snow or spattered in any degree with its melting remnants, but the burgeoning time has come back to the forest.

It is only a half-dozen weeks until the longest day of the year, when false dawn dimly lights the clearing at 2:30 in the morning, and the long twilight does not vanish from the west until after ten the next night. The sun rises far to the northeast, swings around a high arc like a drawn bow, and disappears behind the northwestern hills.

The green-and-blue days of the north woods summer are familiar, not only to vacationers, but to anyone who watches television or goes to the movies or reads magazines or books. Fewer people know the very early dawns.

The first light reveals the trees and brush as greenish-gray shapes, like amorphous, primordial predecessors to themselves. Across the lake, mist rises and coils above a marsh, ghostly and smoke-white as the light increases and the green growth loses its formless mystery. Somewhere a bird chirps, a squirrel sputters into lively argument with itself, gulls pass like shadows overhead, and the eastern sky flushes with pink. The rim of the sun lifts over a hill to send a sparkling path across the lake, and the dawn wind rolls the water gently in bands of rose and powder blue.

From many sunsets watched on warm, moist evenings come a few to remember all your days. Ade and I once saw a thunderhead rise in front of the sinking sun, the cloud's face black and ominous and streaked with blue lightning, its top, boiling into the brilliant upper air, rimmed with fire. Slowly it moved eastward, gray rain veiling the hills beneath it. As it passed north of us, rain swept below the still rising mass in violet draperies thousands of feet high, swaying and folding as the thundering cloud pushed steadily ahead, its bottom in shadow, its sides dull red, and its top turned to molten metal by the upward slanting rays of the sun. Ahead of the storm, the sky was pale green and the hills were indigo. Our lake lay quietly unaffected, turning slowly from mauve to gray. Behind the cloud, a rainbow formed, with a soft ghost of itself below. This double bridge followed the storm until darkness dissolved it.

The strength of the forest is measured in water, and its safety— and ours—depends on rain. Our only frightening days come in a spring when the land is clear of snow and dry before the ice goes out. The dehydrated trees and brush, not yet moistened by rising sap, and the wind-dried duff, are invitations to forest fire. The ice is too weak to bear our weight and the water is so cold that we could not endure immersion in it. We would be trapped if fire should break out between us and the road to town. In the old days, our chances would have been negligible. Today we would have a grim wait for rescue by a Forest Service helicopter.

Every time the sky clouds over during dry periods, I listen for the sound of rain. Often it taps in a transient shower that does no more than strike rings in the dust. Sometimes it gushes in torrents that rush across the baked earth and drain into the lake, leaving only deceptive puddles that evaporate in the sun. Then, after a day when the milky lake has rolled steadily from the northeast, there is a pattering on the roof as of the feet of many mice. It quickens and steadies to a rumble—and the soaking rain has come.

I go out to feel the cool drops on my face. I watch the water darken and soften the dun-colored soil. Drying my wet hair, I look from the cabin window to see the branches bend and drip under the weight of the raindrops. Trickling moisture blackens the bark of the cedars and the pines. The layer of fallen needles beneath them takes on a deeper brown as it soaks up the life-giving water and lets it penetrate to the forest floor and on down to the waiting roots.

In 1961, when our rain gauge showed only a half-inch of precipitation during the six weeks preceding September, many of the grasses and shallow-rooted plants dried up and the tops of the big trees were drooping. Then Hurricane Carla, passing far away on a northeasterly course, brought eight inches of rain—five-and-a-half inches in ten hours. The lake filled like a bathtub, while Ade and I cheerfully put pots and pans under drips where the rain had driven through our supposedly tight roof.

Three days later we had a second spring. Grass and buttercups and trilliums put up new shoots and, when the deciduous leaves began to turn yellow, dandelions showed out-of-season gold on the ground. The mushroom crop was very scanty, though, and the rains had come too late to help the evergreens produce their cones. The chipmunks and mice and squirrels would find the winter ahead a very hungry one, for no artificial feeding gives them a proper diet.

Autumn begins here in September, heralded not by flaring banners of color, but by the "fall sog." The sky is overcast; the air is chilly and windless; foggy ghosts, coiling downhill to drown themselves in the lake, eerily seem to pass through unresisting tree boles; drizzle brings damp that would make a frog rheumatic. Ade lights our first fire and its smoke falls wearily to the ground, joining with the dampness to add a smell of sodden burning to the dispiriting atmosphere. The exodus of summer people begins.

The wild things are in a fever of prewinter activity. I dump the feathers of a defunct pillow into a box in the woodshed and add a few newspapers for the convenience of nest refurbishers.

A squirrel investigates, then pulls and scratches and yanks mightily until she has detached a quarter-page of newsprint. She grasps its edge in her mouth and tries to push it ahead of her, but it traps her hurrying hind feet. She and the paper do a forward somersault off the woodshed floor onto the ground. She picks herself up, chattering anxiously, and examines her tail, which seems to have been twisted in the fall. Tail all right, she turns again to the paper, now somewhat crumpled. A wild skirmish ensues, in which the paper seems to take on a life of its own. Gradually it is torn and wadded until the squirrel discovers that she can cram it into her mouth and, with caution, hop across the yard. At the foot of a black spruce she pats the paper into a tighter ball, squeaking and chirping the while, before she begins the seventy-foot climb up the rough bark. The paper escapes from her teeth when she is a third of the way up, and down they come, the paper to sail off into some maple brush and the squirrel to land flat and sprawled on the duff. She sits up, paws across her chest, and, stamping her hind feet, complains vigorously about such undeserved harassment. That over, she goes after the paper, wads it up again, and this time makes it to the bough that bears her nest. She looks fearfully down at me as I walk toward the tree to watch, but deciding, perhaps, that I am not a climbing creature, she carefully pats and crushes and pushes the paper into place.

Wondering how she knows that newspaper is fine insulation, I start for the cabin, but stop as a chipmunk, cheeks bulging with corn and mouth full of old-pillow feathers, skids to a stop in front of me. I stand perfectly still as she looks from one foot to the other and up, up, up to the top of my head. Reassured, she jumps between the feet of the colossus and scampers to the stone pile that protects her burrow entrance.

The chippy reminds me that I must send out the order for our own winter groceries before the freight stops its summer-only delivery. From the door, Ade is watching the squirrel as she begins a second assault on the obstreperous newspaper. "I guess it's time to seal up the roof," he says. He gets a ladder, and I dig the grocery catalog out of my file.

Then a chilling wind clears the sky. The deciduous leaves, their work done, spiral down. Frost turns the fern fronds to buff and blackens summer flowers. Showers of pale-brown pine needles trim the roof and weave intricate carpets on the paths, hang like fringe on the fences around the abandoned garden, and throw a network over the dwindling squash plants. Pine cones thump on the roof and fall to the ground. Cedar leaves the color of iron ore thicken the layer of duff. When the wind's task is completed, the trees are clear green again. The seeds of a bull thistle float high on fall air whose tang creeps into the blood, to stir and ripple and emerge in an effervescence of delight.

There are many wings over the water as the flocks of ducks gather and the young ones practice takeoffs and landings. Often one, not quite strong enough, runs on the surface, trying to lift with its brothers. Perhaps it will fly south with them, perhaps not. If it cannot migrate, it will die. Nature does not coddle the weak.

The ducks disappear, the geese fly over, and the green hills across the lake wear crests and brocades of gold. Ade and I, if our work allows leisure time, cross the huge rocks that guard our shore and take a boat ride. As we paddle out from the skid, we admire our mountain ash, its leaves a weaving of pale-yellow daggers against which its scarlet berries hang over the water, The gray jays are snatching the fruits, dropping many of them. The spilled berries bleed on the stones or disappear in little splashing fountains. There is movement under the water but the rippled surface is too deeply shadowed for us to see what lake dweller is feeding on such beautiful food.

The arc of the sun is shortening for winter and our south shore is in shadow. On the edge of the lake, maples flame scarlet and alders modestly display foliage of dusty rose. The smooth water is black-green, reflecting the autumn trees like searchlights turned down into the depths—a brightly colored inversion of the northern lights. Among the evergreens, the birches and aspens show pale green in sheltered spots, buttercup gold in the open, shining bronze where the big ones rise above young spruces, and rusty on the ridges where the leaves wait to fall in the wind.

The shadows of the past lie on the hills. Where glacial till alternates with ridges of rock, wide yellow swaths follow the soil from the water to the hilltops, and stunted cedars cling by twisted roots to crevices in the stone. Nearby stands a climax forest, grown through centuries and many cycles of grasses, shrubs, and various trees to full maturity. The stubby limbs of its black spruces, rising high above the aspens, are so thickly needled that the trees are top-heavy. Behind them, surviving white pines remember their fallen brothers. In the distance a saw-toothed ripple bears witness to an ancient folding of the earth. The nearer hills, where evergreens and red deciduous treetops mark the slopes with hieroglyphic designs, are veiled ever so lightly with blue, while those more distant grow fainter row on row in a purple haze, the worn-down roots of the mountains.

We leave behind us the last of the lodges and cabins, and idle beneath granite bluffs, splotched with the maroon and green of lichens. In a little bay we come upon a beaver house. One beaver swims slowly across the water, its head leaving a rippled streak behind. We think we are unseen but, with a splash and a scattering of drops, its flat tail slaps the water in warning. We marvel at the swift, water-skimming flight of a pair of late goldeneyes. We paddle carefully through shallows where the jagged black teeth of the earth snarl up at us through the water. We stretch our legs on a secluded beach, where a stream pours out of a green tunnel.

The sun is low in the northwest. Ade starts the outboard and turns homeward over water that is flat and milky-blue in the evening calm. I watch the waves curling away from the boat in moving sculpture to collapse in a flurry of bubbles at the sides of our wake. At first the waves are blue with silver arabesques flowing over them, their bubbles clear and rainbow glinting, their scattered drops white like pearls or touched by shadow into spheres of blackness. Our trail upon the water lies behind us, gradually turning from turquoise to green. The slanting light is darkening from white to saffron to bronze, and details of the shore stand out in amber-traced clarity. The waves are green now, their patterns and droplets glowing like melted copper. The light begins to fade rapidly, and the waves are violet meshed with restless black.

In the west, a cloudbank is forming, rayed from its flaming edges by the hidden sun. A wind is rising and there is a chill in the air. Far ahead of us whitecaps flash across the ultramarine water like leaping fish. Ade opens the throttle and we race the squall home.

As we haul the boat to safety on its cedar-log skid, the wind reaches across the lake, beating the rolling water into foam-edged combers. Thunderheads are thrusting into a third of the sky and their voices rumble in the distance as the first lightning streaks over the forest. We run to the log cabin, ducking our heads against puffs of dusty, twig-laden air.

Unmindful of the weather, the tame squirrels are gathered in the yard for an evening snack after their long day of cone harvesting. We hand them graham crackers, while chickadees and woodpeckers, blue and gray jays, pick suet from feeders and snatch corn and crackers from the ground.

Suddenly the storm, cloaked in premature darkness, roars toward us through the trees and we and the wild things take cover. Inside, by the mellow light of oil lamps, we listen to the strange voices in the wind, to the crash and thud of waves breaking on the rocks. We hear the crack-crack-crack snap of a big tree falling nearby. Ade blows out all the lamps except those near us, which we can extinguish quickly. If a tree should strike the house, we do not want fire to add horror to confusion. Torrents drum on the roof and pour from the eaves; beyond the black windows, the forest appears and reappears in lightning-white flares; thunder blasts and reverberates between hills and clouds.

Then it is over. The sky clears from the west. Water drops glimmer softly along the eaves as the twilight fades into night. It is time to light all the lamps, draw the curtains against the big dark, eat, and reopen the books we laid down the night before.

It is a good idea to look around before leaving the house on these last warmish days when wild food is fast disappearing. Once I surprised a bear, contemplating the removal of a suet feeder, a nourishing bite before hibernation. He reared taller than a man; with his forelegs lifted and half-spread, and his body thickened by his heavy winter coat, he looked as broad as a door. I froze and he froze. After a moment's consideration we each turned around and went our separate ways.

In November, a steady, purposeful wind strips the last leaves and brings cold rain. The air is raw. Ears ache and hands chap. The chipmunks are tucked into their burrows until spring, and damp, mussed squirrels gnaw bits of heat-producing suet through the mesh of the feeders. The blue and gray jays spend more time around the house. The crows have gone, leaving a silence to be filled by the *crawnk-crawnk* of the ravens. While Ade and I are reading at night, enjoying the snugness of snapping logs and lamplight gleam on the handhewn walls, we hear a faint rattling against the north windows. In a beam of lantern light from the door we see shining rain drops, bouncing balls of sleet, and, here and there, a big, soggy flake of snow. We settle again in our big chairs. The house is tight; the fuel and food are stored. Let the winter come.

The advancing first snow hides the hills across the lake so that from our shore we look north into nothingness. Gradually the fine flakes engulf the house and climb the hill on the south, veiling everything in a semitransparency that is alive with the slanting movements of its particles. Here a flat stone, and the broad dead leaf of a plantain, turn white. Little tufts form in the crotches of the willow branches. A mass of tan reeds becomes an angular design in three dimensions as the white collects, here on the leaves, there on the ground beneath. The mass of the trees begins to take on light and pattern as each twig is thinly covered. In a few hours the dim world is changed to one of clarity. Tree trunks are sharply delineated by thick spatterings of snow on their north sides. The ground is inches deep in white, but still has the snowbent tops of tall weeds to break a smoothness not yet marked by tracks. The tips of balsam and spruce branches are turned into white-feathered ptarmigan feet.

Arctic air pushes south after the snowstorm and we wake to a temperature of around ten below zero, and to the short-lived miracle of hoarfrost on the land. The cedar leaves are edged with spiderweb lace; the branches and trunks, the weed tops and the garden fence wire, are trimmed with fragile feathers; long frost fronds hang from the eaves, like the antennae of giant moths peering over the roof edge. The clusters of green on the great pines by the cabin are balled with frost so that the trees are a giant's orchard in full bloom. The bare branches of a birch on a hilltop are covered with glassy needle shapes. The air has been very still to allow the formation of this ice, so frail that the breath of my passing shatters it, so fine that the heat from my skin melts it as I bend near.

With the day's temperature rise, the evanescent decorations vanish. The giant's fruit trees catch a rosy light, then slowly turn into pines again. The bare birch is hazed with vapor from the sublimation of its temporary silver hair, and faint rainbows flit among the twigs as the mist catches sunbeams through the branches. A slight and ceaseless shower of glittering frost, moisture freezing out of the air, falls like snow from the clear sky, and a shimmering layer of air waves lies above the lake as the great volume of water gives up its heat.

The freezing spray has glazed the shore rocks by December and, in the bays, there is windowpane ice that snaps and tinkles with the music of glass wind chimes. On a quiet evening we hear the improbable sound of barking seals. The ice along the shore, now an inch think, has broken into plates, whose edges grind together and make the seal noises. The small floes pile up and the freezing goes on, until the lake is white and still.

This freeze-up pattern is one of our most regular seasonal effects but, as with all the happenings that depend on this land's capricious weather, the unexpected may wait just beyond the hills. In November 1960, the thermometer dropped below zero and stayed there, day and night, while the wind pushed continually across the lake from the northwest. When the icy air reached our shore, it deposited as rime the moisture it had picked up from the rolling open water, and built up a panorama of splendor that would be a once-in-a-lifetime sight even for a native of the north.

At first there was a light tufting of crystals on branches and twigs, tree boles, and rocks. The layer grew day by day until the twigs were inch-round and heavy with white, the tree boles wore vertical flutings, and the rocks lost their individual forms under their thickening cover. The wind still blew and the turbulent lake did not freeze over. Two weeks later, the tree trunks were frosted eight inches thick on their north sides; the living branches, their original intricacy lost under glittering padding, drooped like those of the weeping willow; dead and brittle limbs were broken off by the weight of myriad tiny crystals. Bushes collapsed into alabaster heaps, and the rocks along the shore vanished entirely in a bank of smooth and shining whiteness.

When the bitter wind finally died, I went out on a point of land and looked back along the shoreline. Except for the tops of the big trees in the background, the familiar forest had vanished. I saw only masses of white. Forty-foot balsam firs bent under balls of frost half as wide as the trees were tall, their branches and frost coverings interlaced from tree to tree. Everywhere the shore was white, not the sparkling, unbroken white of fresh snow, but a whiteness of form and design, where pale flowers sprouted from ghostly earth and rippling, folded shapes were outlined by the palest shadows. For so pure was the whiteness that even the shadows seemed to be of a lesser whiteness than the unshadowed places. I was looking at a mass of living light, not bright with the touch of the sun and dark with the accent of shade, but made up of different degrees of whiteness, of different textures of light.

There was a tinkling as of little bells near my feet and I looked down at a thin sheet of ice along the edge of the water. The supercooled waters of the lake, at last released from their restless rolling before the wind, were starting to freeze.

As I watched, the ice sheet spread out from the shore, at first thin and rippling like transparent silk, then thickening and stiffening. Other sheets were reaching from the far shore and forming in the open water between. They spread and rushed toward each other, stirring the surface ripples in blue floods over their edges, there to freeze on top of the growing sheets. The movements of the great, flat planes, stretching like disconnected tiles over the lake surface, a half-mile wide where I stood and several miles from east to west, brought flashes of sunlight, as though from acre-sized mirrors. The lake, its surface as smooth as the stillest pond, gave an impression of heaving with the geometric patterns in a giant's kaleidoscope.

And then the appearance of movement grew less. The blue areas of flooding water vanished. The mirror flashes were gone. The sheets of ice had joined and sealed the miles of water for the winter to come. The whole marvelous change had taken place in little more than an hour.

Even as I watched, hoarfrost began to form in little buds on the clear, new ice. During the next three days the frost on the shore began to sublime into the now dry air and the moisture was redeposited on the lake's ice sheet. The frost buds grew—into acres of crocuslike "flowers" and finally into frost "roses" as big as cabbages, their frail petals made of the most delicate and ephemeral crystals, so fragilely connected that the slightest zephyr would have destroyed them. In the below-zero cold, the warmth of the low sun did not affect them and they glittered amid their own blue shadows. Then came clouds and snow, and the frost roses were forever gone.

When the air temperature drops far below zero at night the ice thickens rapidly, especially if no heavy snows insulate it. The expansion of the solidifying water sets up enormous stresses that cause cracking, sometimes from shore to shore. Once when I was standing on the ice I heard a whining like the ricochet of a bullet. I felt a bump underneath me. While the whining fled away to end in a snap like the crack of a bullwhip, I looked down to see an inch-wide crevice between my feet, from which water was flooding around my boots.

When the ice is a foot or more thick, its cracking fills the air of bitter nights with sounds like the crashing of thunder and the slamming of big guns. Minute internal cracks and slippages join and lengthen, weakening the whole frozen layer. Eventually the ice breaks; there is a rumbling that grows louder as the crack approaches our shore. When the shock of the icequake is transmitted through the bedrock of the land, we feel a heavy jarring as the underground shock wave lifts and lowers the cabin. The wave continues with lessening vigor until it dies out in the rock layer, its initial strength and the density of the rock determining the distance of its travel.

The ice on one side of such a break usually lifts above the original ice level, throwing up a pressure ridge that is sometimes several miles long. The risen layer may be only a few inches high, producing a sort of curbing across the lake, or it may rise several feet, extending over the lower edge and leaving open water with broken chunks like little icebergs.

The blue-green of this ice puts to shame that anemic shade that fashion designers call "ice blue." Its blue has the brightness of summer skies and its green has the richness of reflected moss and ferns. The very bubbles of the brooks and the foam of rolling waves are caught within it, in chains and clusters that the imagination can turn into the ghost forms of flowers and leaves. The underside is covered with glasslike dendrites, faithfully reproducing the branching of trees. "Blue ice" holds the shadow of summer past and the promise of summer to come.

Such ice can support cars and even heavy trucks, but it presents danger that is the greater because it is not obvious. The pressure ridges are not readily seen when they are approached from the high side, especially when they are so aligned that they cast little shadow. A snowmobile or snow sled could bump off a low one with slight or no damage, but if it should run over the edge of a high ridge into the open or thinly ice-skimmed water below, it would vanish into the maw of the lake.

A "flowage," the local word for a series of small lakes connected by streams, has a strong current that flows throughout in one general direction, but that changes with channel depth and width. The ice along the channel's sides is constantly freezing and thawing under the influence of the moving water. This varies the channel size and creates unstable currents that affect the surface ice. On one day the flowage may be solidly frozen over. On the next, a current may have undercut a section so that a breather hole has formed, surrounded by a slush of flooded snow that is hidden under a newly formed thin layer of top ice. Flowage ice is never safe. Oldtimers test it carefully, pounding ahead with poles that are strong enough to bear their weight and long enough to cross the opening if they should break through. Strangers may look on a flowage as a good trail for a snow sled and mistake breather holes for ice fishermen's locations, often with chilly results.

Large lakes have underice currents, too, most often near narrows or islands. They produce breather holes and thin ice, the latter of which is the more dangerous because its thinness cannot be seen. Travel on ice, especially in motorized vehicles, which are heavy, and on snowshoes, which are clumsy, should be undertaken with caution.

By the first of the year the roads wait for the snowplow and air from the tundra brings deep winter. Now the whine of a powersaw three miles away may sound as though it were next door, and the dry air lets the morning's forty-below temperature rise to five above at noon. Although the shortest day is past, the sun is above the true horizon only from 7:30 to 4:30, and is so far to the south that it does not appear above the treetops. Only a few beams reach our windows and much of our light is reflected from the snow. The sunsets on the clear, cold afternoons are far to the southwest, but all the horizon is ringed with red that fades slowly to coppery rose and then into pearly dusk.

After a midwinter snowfall, the forest has the picture-card look familiar to everyone, but to be inside that picture is a far different sensation from looking at its printed representation. All around, in a windless silence, the snow spreads and flows and ripples, ermine-smooth under clouds and diamond-bright in the sun. There is a white-Christmas feeling that is unrelated to philosophies or carols, to merrymaking or gifts: a feeling that endures the rumble of SAC jets and is undisturbed by threatening tomorrows. Maybe this is the "peace that passeth understanding," the happiness that we humans pursue so savagely among material things.

The moonlight, through the dry and dustless air, is so bright that even a crescent brings shadows to life. I stood on the lake ice one afternoon, bemused by the sight of gray sun shadows slanting from the west, crossed by purple full-moon shadows from the east. After the twilight fades, the moonlight turns the icicles on the eaves to translucent blue fringes and strikes sparks from the frost platelets on the snow. The clearing is mystical and strange until I see movement at the edge of a patch of shadow, and make out our tame doe, patiently waiting for some protecting darkness to cover her feeding place.

Unless the sky is heavily clouded there is always some light in the winter night—the bluish aura of the moon, the low, green glow of the aurora, the scattered pinpoints of the stars. They blaze against the clear, black sky and sometimes shine through the mistiness of the auroralike jewels on the veil of Time. The red and blue giants gleam pink and azure amid the twinkling whiteness of their companions, and the planets carry steadily glowing lanterns along their ancient paths.

The aurora shows a dim light low in the northern sky all year, but reaches its splendor in winter when dry air and long hours of darkness make way for its fabulous light. Sometimes it flames red as the hearth fires of Valhalla or green as the waves that bore the Vikings' ships. Or its cold-white banners and curtains strengthen into searchlights that set the birds twittering. Once I saw a squirrel and two chickadees come looking for breakfast by aurora light, only to depart in confusion when the "dawn" faded from the sky.

There was a night when I woke to windows filled with a pulsing, unearthly green. Hastily wrapping myself in clothes heavy enough to defeat the below-zero air, I went out into a night that one might be more likely to encounter above the Arctic Circle. The wild green flare lighted my way as I skittered down the hard-packed snow of the path onto the wind-ruffled drifts that covered the lake ice.

A mile to the north, the Canadian hills crouched low before a cold and flaming glory. Two silver arcs towered above the rolling horizon like a faded double rainbow, a shimmering band of platinum between, a crystal-flecked velvet darkness beneath. Rays of white and rose swayed high above the arcs and the wavering green overhead seemed to come from everywhere. And, in the quiet night, my ears caught the faint swish and rustle, like lightly touching taffeta ribbons, that is the voice of the aurora.

A distant hum came from the west and grew into the throbbing roar of an airliner, approaching on an emergency route. It passed directly overhead, flying low, sleek as a fish, gay with its red and green and white lights. I wondered whether anyone beside its crew was watching the north with me. The plane flew out of sight, an incongruous reminder of a civilization so much more remote than mere distance could make it.

The lights in the sky grew dim. The platinum slowly lost its sheen and the lower darkness grew larger and deeper and seemed to come terrifyingly nearer—a black whirlpool, utterly without light and of endless depth. As my eyes tried to penetrate this opening into space, the northern lights faded. Only a faint glow lingered to outline the hills, and the gray distances stretched on and on.

MEET MY PSYCHIATRIST

Les Blacklock

Tense? Frustrated? Discouraged? Have a tough decision to make? This may be the time to visit my friendly psychiatrist, Old Doc Log. You'll find him patient, willing to listen. And his price is right. His procedures are beautifully simple; the results, impressive.

Occasionally, routine jobs accumulate seemingly beyond our ability to cope with them. Catching up seems impossible. When that happens to me, when I've had it up to *here*, I head for the nearest patch of wild. Doc Log has branch offices wherever there's a rock to sit on, a tree to lean against, waves lapping a shore, a rippling stream. Even a park bench can bring results if there are trees, squirrels, pigeons, ducks—wild, natural things instead of the hard, square lines of man's world.

When I was a student at the University of Minnesota and homesick for my native north woods, my psychiatrist was the bench in front of the whitetail deer habitat group at the Museum of Natural History.

Whenever possible though, I like to visit Old Doc Log himself. He's an old friend as well as my calmer-downer, and a session with him is as enjoyable as it is helpful.

On a recent visit, as I drove up the wild road toward Doc's office, the roar of the engine and the keyed-up tenseness of a long drive kept me from remembering that I would experience *silence* as soon as I stopped and turned off the engine. When I did stop, I was so awed by the quiet that the click of the car-door latch was disturbingly loud.

It had been several months since I had been beyond the constant sounds of civilization—freeway traffic, jet planes, power mowers, sirens, air conditioning systems, refrigerators, telephones, TV, people—with never a moment when all were still at one time. Now I could not hear one of those sounds, nor any sound. It was like relaxing music, a symphony of silence.

There is no trail to Doc's office, but it's just off the north edge of a grove of ancient white pines, easy to find.

I walked across the road and stepped into the shallow ditch. A loud WHIRRR stopped me as a ruffed grouse flew from the far side of a thick clump of balsams. I followed its flight by the ticking of its wingtips against branches but failed to get a glimpse of my favorite bird.

"Slow down, Les," I told myself, "and be quiet." Then I started a very slow step/listen/step/look pace. I wanted to *see* the next critter I met.

It worked. A snowshoe hare crouched quietly in its "form" until I was within fifteen feet of it. Then it raised a bit and rocked forward and back, "cranking up" to leap away. I spoke quietly to it, and it froze. I was in midstride, so I softly lowered my foot as I talked, waited a bit, then slowly started another step. That was too much. As it jumped off, I whistled softly, and it stopped and sat up on its haunches to see me better. I talked to it for several beautiful minutes while we studied each other. Its nose twitched, and I almost expected a Disney-cartoon voice to answer me.

Under the pines I joined a deer trail and followed it as long as it was going my way. There were many fresh hoofprints in the duff, so I was especially quiet, or tried to be. Heel down slo-o-wly, then ease forward onto the entire foot, trying to compress the needles and unavoidable twiglets so carefully that they wouldn't snap.

"Jay! Jay! Jay!" My presence was announced loudly by a bluejay, so my soft-pedaling stalk then seemed rather silly.

At the north edge of the grove I entered a thicket of head-high greenery. There was a brief flurry from a low cluster of vegetation, and a white-throated sparrow, disturbed and anxious, flitted to a nearby mountain maple. I carefully avoided her nest site and circled around to where Old Doc was waiting.

There he was, larger and more impressive than I had remembered. I stood quietly, admiring the huge moss-covered trunk surrounded by lush ferns and wild flowers.

Old Doc is a model of exemplary living; he took what he needed from the earth, but he was given back much more. For three centuries he stood here, recycling soil, water, and air. And even now his life starts to flow into new life—mosses, lichens, and fungi. In time his body will be soil, and he will become bunchberry and fern, clintonia and wild rose, birch, balsam, and spruce.

I walked over to Doc, patted his sturdy flank and sat down on his moss-cushioned couch.

I don't really talk to Doc. Not out loud. Our communication is sort of by osmosis. But as I sit there and let it all out, somehow my lesser problems seem pretty inconsequential. Major problems are still major; but as Doc patiently "listens," I can sense what his "answer" is going to be. He waits quietly while I ramble on about a problem that's really been a stickler. Then in the clear silence he allows me unlimited time to mull over the alternatives. It finally dawns on me that it's my move, that Doc is waiting for *my* answer.

Surrounded by the beauty and wisdom of nature, my problem somehow seems less formidable. In this uncluttered bit of time it has been possible to see it alone, in new perspective. My body and brain are refreshed.

I get up.

"Well, Doc, if we're going to lick this thing, I'd better get at it!"

And I do.

SNOWCAMP
ROMANCING WINTER IN THE LAKE SUPERIOR WILDERNESS

Mark Sakry

*Fifty degrees below zero was to him just
precisely fifty degrees below zero. That there should
be anything more to it than that was a thought that
never entered his head. . . . The trouble with him
was that he was without imagination.*

—From "To Build a Fire" by Jack London

To know winter in the wilderness is to know silence. When all is
still—when the arctic storm has dwindled to a sighing memory
and the forested lakes stretch unbroken before you with a fresh
cover of new-fallen snow—the silence is overwhelming. You cannot really hear it, as some say you can. Yet, for a time, a certain
loudness rings.

Then, listen closely, beyond the silence, and you begin
to hear faint voices. Hushed whisperings seem to emanate from

nowhere and everywhere at once, as though from some in-between world, somewhere between death and rebirth—the sleeping billions: multitudes entombed by ice and snow and infinitesimal cold speaking through the immense silence. Perhaps of an afterlife. Perhaps of the promise of distant spring.

And just when you think you are beginning to understand, a breeze breathes through the trees. The whispers vanish. Snow hisses softly down through the pines as their peaks rock gently against the sky, then the gravelly *grawk-grawk* of a raven on the far shore. A muffled groan rises from the lake as the ice yields to the sudden, gentle tug from above. Somewhere, a pop from a fire and the murmur of voices inside the forest.

This is the setting, the magic. This is what compels a handful of adventuresome souls every winter to venture off established trails into the frozen wilderness of the Lake Superior region.

Some ski, some snowshoe. Many go just for the day, realizing they must return before night casts its mortal shadows across the forest. As many see it, to be caught in the dark would be folly. The moon, if it's full enough, might defer risk for a while. But it rests on one's best reckoning to quit the fading forest or be forced on humankind's most ancient and incumbent measure for surviving the winter night—to build a fire.

Yet some see it differently. Just a few. Those with the wherewithal. And an incredible amount of imagination.

The Inuit had many names for it: *pukak, upsik, anniu.* To the rest of us, it's all the same stuff: snow.

To sleep under it, to repose in sublime comfort among heartfelt comrades inside a hollowed-out mound, while the wind howls outside on one of the coldest nights of the year, is indescribable. To the winter adventurer, it is the height of romance, the consummation of his or her wilderness ability.

Indeed, it is not just imagination but ability which makes accessible some of the most enchanting haunts of the Lake Superior region and fully opens the door to the winter season.

Once one attains the essential skills with which to comfortably endure extended stays in the winter wilderness, larger-scale excursions are possible (not unlike their summer canoe-trip counterparts) into such areas as the Boundary Waters Canoe Area

Wilderness of Minnesota and Quetico Provincial Park of Ontario. Here the combination of scenic forests and flat lake travel affords immeasurable beauty over long distances.

For shorter, though possibly more strenuous, wilderness excursions, Porcupine Mountains Wilderness State Park in Michigan and Cascade River State Park in Minnesota offer miles of rolling forest in pristine alpine settings. Many state and provincial parks in the Lake Superior region offer groomed trails as well as winter camping. All, by virtue of the season itself, afford almost total seclusion and unadulterated winter beauty.

This also applies to other remote areas where, with the exception of snowmobiles, accessibility by mechanized travel is limited during winter months. For instance, winter wayfarers will find suitable conditions for a wilderness snowcamp in many places beyond the road heads of Superior, Chequamegon, Hiawatha, and Ottawa National forests.

Nothing compares with the special confidence gained by facing the winter elements in a snowcamp for an extended stay in the Lake Superior wilderness. Nor the beauty. But the essential skills—knowing not only how to "survive," but (more importantly) how to be comfortable—are necessary before winter becomes less formidable than it was.

If there is a single key to staying comfortable, even in the kind of cold which grips the northern Great Lakes region, it is doing away *completely* with the notion that one must rely on an external heat source, such as a stove or campfire. Winter campers depend on internal heat for warmth.

The body is viewed as a furnace fueled by high-calorie foods. These are rarely taken in the form of sugar; they are eaten (in seemingly gluttonous quantities) in the form of fat, protein, and complex carbohydrates, like nuts and whole-grain cereals, which burn more slowly and substantially. The furnace is then insulated properly—no cotton, please—in layers of clothing that may be added or removed to suit the level of activity. This prevents *over*heating and excess perspiration, which can dampen clothing (especially cotton), ultimately causing heat to be lost faster than the body can generate it. The condition is known as hypothermia.

Plenty of fluids (more than you need) are routinely consumed to prevent dehydration common to winter activity. Body fluids must be maintained to properly metabolize food fuel into heat. But no alcohol; this promotes heat loss. Coffee is avoided because it is diuretic; it depletes body fluids.

Water is obtained by melting snow or boring through lake ice with an auger and dipping it out. While many areas in the Lake Superior region provide pure natural water, it is still good practice to boil water before consuming it.

To be truthful, overnight shelter, even in an established snowcamp, is unnecessary. That is, if you have an adequate sleeping system consisting of a fully rated winter sleeping bag or combination three-season bag and liner, with a closed-cel foam pad between your bag and the ground. Otherwise, a *quin-zhee* (pronounced "kwahn-chee") snow shelter, fashioned by hollowing out a mound of snow, will keep you surprisingly warm, even in a three-season bag, if the shelter is built with its floor elevated above the top of a small door opening. A tent will keep the wind and snow off, as well as help consolidate your sleeping area, but it won't insulate you from outside temperatures.

The pack sled, or pulk, has rendered the backpack nearly obsolete for winter travel, especially over flat lake country. It is not only easier to tote, it extends the amount of camp hardware and equipment you are able to bring along (about the same as a summer canoe trip). With a harness you can put your dog to work, too.

February is a good time to plan an extended trek. Average daytime high temperatures around the Lake Superior region are around twenty degrees above zero. But don't be caught off guard. It can still hit subzero; best to be prepared for the worst. Plan all winter treks as though you were going in mid-January.

And if you've never spent an entire night outdoors in winter before, try it in your back yard first. It's your safest testing ground.

Wilderness was once defined as that area of undisturbed natural land resting beyond the road head. Truly, by that definition, winter brings to this region, already rich and abundant with wild land, a wilderness of immense proportion. Indeed, the stillness invoked by Nature's hand upon the vast forested areas about Lake Superior amplifies the implicit solitude and romance of the region.

It is accessible to anyone with a pair of skis or snow-shoes, and the wherewithal to negotiate even a half-day trek beyond the trail head. But with a little more imagination, you may be compelled to advance your sensibilities beyond the perimeters of daylight into the unvisited realm—of the sleeping billions.

It waits. Still. Silent. Snowcamp.

PART TWO
INTERACTION

BASSWOOD LAKE

Heart Warrior Chosa

Every summer, all summer long, we went to what is now known as the Boundary Waters Canoe Area in the furthermost northern Minnesota-Canadian border. It was not a national park then but privately owned, with many resorts for rich people, besides our modest cabins.

We spent the most wonderful time of my life there. It was as beautiful and warm as the winter in the city was cold and starving. Our apartment was sparsely decorated and our food meager in the city, but when we went up north to lake country in the summer, everything changed. We rode the old milk train there and sometimes sneaked our cat in a lunch basket aboard with us.

Daddy went ahead to get the cabin ready. The cabins were built during the Depression by hired Finlanders and belonged to Grandpa and Grandma. In the late 1800s northern Minnesota boasted logging camps and the foreign settlers mainly came from Yugoslavia and Finland with a small sprinkling of

Italians. They settled in a town south of the BWCA called Ely, Minnesota, and were decent folk who learned about woodland life from my grandparents and other Indian families from Burntside and Shagawa Lake.

There were three cabins on the island which all the family shared. Over in Hoist Bay there were more cabins they shared. We always stayed in the middle cabin on Woodpecker Point.

The cabins were crafted by masters of such work, the Finlanders being famous for fine craftsmanship in wood. They were made from cedar logs fit snugly together so chinking was unnecessary. Each log was the same size and length.

Our cabin had a screened-in porch. A huge, stuffed leather rocking chair stood on the porch, and a screen door led to the cabin from there. There was a front door that faced the woods. On the porch side was a view of the lake. That side facing the lake was up on stilts since a sharp drop led down to the water's edge. In one corner of the cabin was a woodburning cookstove, a table, and dresser. Kitty-corner was a double bed with a mosquito net over it, which my brother Tommy and I had fun falling into.

Dad stayed at the resort during the week, where he guided rich people fishing. Sometimes he'd get one-hundred-dollar tips. Mama took to the country like a long-lost waif finally returning home. She soon became an expert fisherwoman and confounded the professional guides with her catches. We spent many a sunlit day hauling in fish on our homemade fishing sticks. Some days we went by boat to a huge sand beach for swimming and picnicking. Later, I found out it was the favorite village spot for my ancestors.

During the summers from ages one to five, we ate from nature and lived the good life and were in the pure waters being healed.

There were several paths that led away from our cabin. The first one, very steep and with many rocks led down to the shore directly in front of the cabin. The water's edge was teaming with leopard frogs who had caves along the shore. There were many fish, snakes that swam in pairs, and painted turtles basking in the sun on an old stump. Butterflies and dragonflies swam in the air, and wee toads hopped on the dusty earth, along with the busy ants and other crawling, hopping critters.

There are no more frogs in these waters, because in the late 1960s and early 1970s they came and got all the frogs so white children could dissect them in biology classes across the country. What a deprived way to meet a frog. Consequently, the beautiful water snakes are gone, too; they lived off the frogs. But when I was little, they were everywhere.

In fact, I used to think they were sick because their tummies were so white. Then I'd capture one at a time and take them down to the beach, on another path that led from our cabin, led through the tall pines upon a cushioned red pine-needle floor to the end of the island we were on. A clearing opened on a sandy slope to the sand beach.

Once there, I hunted up a flat rock and turned the frog upside down, holding tightly to his little squirming arms and legs, waited patiently until he got a good tan, but he never did and seemed to grow dizzy. Alarmed, I'd throw him back into the water. Finally, it dawned on me that frogs weren't supposed to have brown tummies, as I watched anxiously for any sign of life. The frog lay dazed for a while, floating on top of the water, then suddenly awakened and swam away, his little legs going fast. He soon disappeared into a cave along the edge of the shore. My imagination spun stories on what it would be like to live there. The frogs didn't mind when the water rushed in on windy days, for they were at home on land or water.

I asked Mama, "How come frogs have white tummies?" Instead of a direct answer, she told me another story. "Judy, go down and lie along the shore and see if you can see some tadpoles." "What's a tadpole?" I asked. "That's a baby frog before he becomes a frog. He has no legs then and can't go on land until he gets his legs." Hmmmmmm, that was interesting, I thought.

One summer's day Mama, Tommy, and I were at the beach. There was also a long dock there built out onto the water for landing boats. Tommy and I were very little, two and three years old. We were laughing and running back and forth. Then Tommy in his excitement ran off the end of the dock. Mama jumped up and walked all the way in under the water and pulled him out. He was all right because he was holding his breath under water and didn't get any in his lungs.

Mama was not a fat woman at all, but had very heavy bones. They were so heavy that she could not float at all. Many times when she took us to Washington Beach, where we picnicked and swam, she thrashed in the water, sounding like a locomotive. Try as she would she could not swim because of the heaviness of her bones. She sank like a rock.

Washington Beach was the largest sand beach on Basswood and several miles by water from our cabin. It was the favorite village site for my ancestors.

One summer when I was real tiny, going on two years old, I sat in an ant pile and when Mama saw me she screamed and immersed me in the lake. The ants were not biting me but she thought they might.

I crawled around on the soft, red pine-needled floor of the forest, looking up at the sky with swaying pine trees above me, their arms rustling and quietly whistling in the wind.

That whole area is one of the heaviest sarsaparilla fields I've seen anywhere in that lake woodland country. At that time, they appeared really big, like trees. I'd sit underneath their leaves and crawl in and out among them, eating their blue-black berries, which looked liked blueberries, but tasted like the smell of gasoline. Mama told me not to eat them because they were poison. I didn't listen, because they were my friends, and I liked the way the berries grew in a cluster circle, all radiating on little stems from a common center stem.

There were hordes of armyworms one summer. They were everywhere, including our porch. I enjoyed playing with them day after day. They were an inch long, smooth, with stripes, and came in assorted colors. I played with box elder bugs the same way in the cities and let them crawl all over me as I watched cloud formations float by, my back warmed by the earth. Sometimes I'd get the uneasy sensation that the earth was moving.

One day I was playing down by the beach, when suddenly I heard, "Buzzzzzzzz." Buzzing in spiraling circles was a big, fuzzy, round, pretty thing that landed on my forehead, on what is called the third eye. I crossed my eyes to see him better, when he stung me.

In pain, I peeled up the path to the cabin, howling all the way. My feelings were hurt worse than the sting because I felt betrayed. "Why did he sting me?" I demanded to know. "Bees do that when they get scared," Mama said, as she went on with another story. "When I was a little girl, I used to catch bees in bottles and try to skin them with a piece of glass. I thought they had such pretty stripes that they would make a beautiful rug for my doll house. But their skins always fell apart, no matter how careful I was."

Another creature that confounded me was the grass-hopper. He clutched onto my skin and for no reason spit brown stuff on me. Mama explained that grasshoppers chewed snuff like Grandpa, and when they spit on me it meant they liked me. But I doubted it, and didn't think they liked me 'cause they spit on me. I continued to play with them, but in a reserved manner.

We had to cross several huge lakes to get to our cabin. The first was Fall Lake. It was very long and took forever to cross, and it was often windy with lake spray wetting my face and roll-ing, churning water caused by the motor on the huge cruiser. The water looked like root beer. Then we crossed over an island on the four-mile portage that was built by my grandpa during the De-pression. It was a narrow dirt road with bumps that rattled the passengers and all their camp gear. We passed dark, murky for-ests whose floors were covered in thick, soft green moss and light, open, birch stands and other little lakes that shone like blue spar-kling jewels. We passed over a little bridge made of a couple of planks over a fast, moving stream, got scared when we went over it, thinking, "What if the driver missed the planks," but he never did.

The end of the portage came out on Hoist Bay and Bass-wood Lake, which is the lake our cabin is on. Basswood is a huge lake and runs for a long way and is divided in half by the U.S.A.-Canadian border. Our cabin is half a mile from Canada by water. Basswood is forty miles long.

At that time many tourists who were very rich, like the Pillsburys and other magnates, used the four-mile portage to get to some plush resort. We all rode together on an ex-war machine painted red and called the Duck. It went on both land and water. It went fast on water, but slow on land. I kept my eyes on the water, as I felt uncomfortable with the other passengers. Some were friendly, but others were snotty and eyed us curiously.

A trading post run by Old Mister Brown awaited us at the end of the portage. He took it over after my Grandpa died. Grandpa had built it during his logging days. Paul Bunyan's grave was alongside the trading post.

Old Mister Brown was a portly, white, balding man and he was always good to us, giving Tommy and me our choice of pop. Huffing and puffing, he told us about the wolf skins, deer skins, fox, and everything on his racks. Or about the time he lived with the Choctaw and Chickasaw Indians, and how nice they were to him, and what a grand time he had there. He showed us his moccasin collection from them and his Indian clothes from the Eskimos, and our own Chippewa.

I never really cared for the pop, but I liked Old Mister Brown and always listened eagerly to his stories. About that time Dad arrived to pick us up and to pick up food supplies. He scolded us and scowled at Old Mister Brown.

Dad proceeded to pick out the food supplies with his pay from the rich people he guided fishing. Old Mister Brown and Dad raced each other adding up the price of groceries, Dad using his head and Old Mister Brown using his machine. Dad always finished first and he was always right. Old Mister Brown always exploded in amazed laughter.

Meanwhile, in the background, the tourists in their shorts, sunglasses, and fishing paraphernalia were milling around with buzzing voices and occasionally peeking at us. Sometimes one of them made a snide remark about dirty Indians, but we ignored them. Other times, while we were playing on our small beach, the tourists went by in small fishing boats, pointing at us, exclaiming, "Look, look, real Indians. See the Indians! See the Indians, the dirty heathens. Thought they were all dead," etc. I shrugged my shoulders in disgust and went about my play. I wasn't going to let them distract me. Besides, they behaved worse than that grasshopper.

Once when we were still real little and sitting on the porch steps coloring pictures, engrossed in our art, Mama walked out on the porch, as mothers will do when their children are very quiet, and looked across the lake. The water lapped gently against the shore. As she turned toward us, she gave a yelp. Hurriedly, Tommy and I turned to see what her frightened eyes were fixed on. There stood a big black bear. "BEAR!" she hollered.

She had already prepared us for this. We had had practice bear drills, just like bomb drills after the war. He was standing up and coming toward us. He had a patch of white on his chest just below the neck. Bears can be very quiet when they want to be. As we were trained, we rushed into the cabin and jumped under the bed as Mama blocked the door. The bear sniffed about on our porch, ate the dog food, and sat in the rocking chair. After a while he ambled down toward the beach.

Mama dug us out from under the bed, as we were huddled there wide-eyed and petrified, and squeezed as far into the corner as possible. Finally, she coaxed us out and went to the porch to see where the bear had gone. She hollered real loud down to the beach, and since the wind was down and the water still, she could be heard at the other end.

After a while an ex-war friend of Grandpa's came over. He had a little shack Grandpa had let him build on our island. He said, "The bear started swimming across the lake. Hurry and get your kids and we can ride out in the boat to get a close look at him. The bear will be so busy swimming, he won't have time to bother us," he reasoned. We all hurried down to the beach where his boat was. We were packing in when Mama noticed the man's gun. "Oh, no! We can't go if you are intending to shoot that bear. He won't bother anyone now that he's going across to the other side," she said. The man promised he wasn't going to shoot the bear and said, "I only want to scare him so he won't come back."

The bear swam a good way across by the time we caught up to him. His nose and ear tips showed above the water. Then a shot rang through the air as the bear disappeared into the watery grave. Only a pool of blood stained the water crimson. I shot the white liar a disgusted glance.

The next summer another incident took place involving a large animal of the forest. Half-asleep and propped up heavily on one elbow, I saw Mama moving around quickly through the filtered white mosquito netting on our big, double bed. The kerosene lamp flickered wildly, casting Mama's giant shadow on the curving log wall. "What's going on?" I asked, still fighting the slumber off my mind. "A wolf's chasing Putsy around the cabin; first Putsy was chasing it, then it turned around and chased him," she whispered urgently, loading the old one-shot gun.

She opened the cabin door, the light shone out into the moonless night, and a loud blast cracked, silencing, if only for a moment, the high-pitched humming of a million mosquitoes that hung in the treetops. Snapping brush, the wolf chased into the forest, swallowed by the black night. Our dog Putsy didn't come out from his hiding place under the cabin for two days after that.

It seemed toward the end of summer that the wild animals became active and more visible. It started out an innocent enough day. We had been to Old Mister Brown's trading post for a few food supplies. Our little motor boat roared across the expansive Hoist Bay, which was always more windy and choppy than other areas of the lake.

Wind-born spray tingled my body, wetting my parched lips, as wisps of hair slashed across my cheeks and the thumping motion of the boat crashed me up and down in my seat. I hung on to the cool metal side of the boat, my hands freezing wet. My pinched eyes squinted in the bright sun, and through narrow slits, jewels of light danced on the foaming crests of waves that rolled against the side of our boat, causing it to rock even harder.

A dark, cool curtain suddenly blocked the light as the boat slowed to a stop in the still waters of one of our favorite fishing spots, the wind-protected narrows. Our boat bobbed, rocking up and down gently in the dark reflection of the tall, vertical black rock cliff. We nestled in close to it, so we could almost touch its shiny wet side. It was deep, cool water; a good place to catch northern or walleye.

Tommy and I let down our lines from our little homemade fishing sticks. The thick black line slid over my right forefinger, causing a red line on it. When the sinker hit bottom, I re-rolled the line a half-foot and waited. Mama cast with her fly rod, her bait landing—kerplunk—sending ripples in circles ever expanding outward as she reeled in her line.

It became habit to fish the narrows next to the cliff on our way home, just in case we might luck out with a quick catch. After a while Mama gunned the motor and we pulled up our line. We passed little bays and islands along the way, sometimes hearing moose battling deep in the forest during rutting season, or in early spring the partridge pounding their chests like ancient drummers of long ago.

Finally, we pulled up to the sand beach, docking there when we had groceries. It was easier to unload, and the path was longer but of gentler incline back to the cabin. Mama led the way with most of the groceries, then Tommy fell in behind her, and I brought up the rear, carrying a box of oatmeal.

I kept my eyes on the ground observing the busy ants scurrying around on their sand hills. Even though it was blowing on the lakes, the thick roof of the forest kept out most light and wind. The air was dry and hot, making the rust-colored pine needle crunch under our weight. The scent of pine sap delicately touched our nostrils, and little birds chirped merrily. Occasionally, a busy chipmunk dropped pine cones on our heads, a favorite trick of theirs that made us laugh.

I abruptly ran into Tommy, who had run into Mama, who stopped. She pointed ahead and to our left just off the path. There, lying in a round depression in a cool, shaded spot, was the longest bull snake we had ever seen. Out of its mouth hung the limp little webbed feet of a leopard frog!

Mama set down her groceries. She looked mad. "Hurry, find some good-sized rocks," she commanded. I set down the oatmeal box, and we all hunted for rocks. I found a big one half-buried in a cool, dark earthen nest with turquoise, stiff lichens growing on its top. Digging around it, loosening it, I handed it to Mama.

She had a little arsenal collected and began to stone the snake. Each time she hit him she grew more determined and pelted him harder with the next strike. I started to feel sorry for that snake. Finally, the long, army-green snake coughed up the frog.

We watched intently as the dazed frog collected his wits and staggered toward the lake. The bruised snake crawled slowly in the opposite direction. "That takes care of that snake," said Mama triumphantly, dusting her hands.

In the excitement, I had backed into a stiff pine branch and broke a hole in the oatmeal box, but didn't know it. Since I brought up the rear of our troop, no one else noticed it either.

That evening as we drifted off to sleep, loons, the state bird, were heard singing in the muted background.

The next morning there was a long line of fat partridges feeding along the path, all the way to the beach. Mama whipped out the ancient and trusty one-shot gun. Loading and reloading, she brought down six of them. Boy, they tasted good—except for the black bits of buckshot we bit into occasionally.

THE DREAM NET

Sigurd Olson

A Chippewa woman gave the dream net to me when I left Grand Portage Indian Reservation on Lake Superior. It was an unusual gift and it pleased me, for Indians do not give their charms lightly to strangers. I examined the little net of fine thread strung tightly on its four-inch hoop of ash, turned it over and over in my hands, wondering wherein its secret lay. It was shaped like a perfect spider's web with a hole in the very center.

She asked me to take it home and hang it over the beds of my children, said it meant good luck and pleasant dreams. When she saw the question in my eyes, she explained how bad dreams as well as good were always in the air around sleeping children, that the bad ones waited for dusk and if there were nothing to stop them a child might scream.

I listened, delighted with the idea, and thought of the times I had awakened in terror as a child, and of the many times my own boys had whimpered and moaned in their sleep.

The Indian woman continued and told me how in the days long before the white man came, a dream net hung in every tepee in the village because mothers wanted their babies to go into the dream world in peace and awaken with quiet eyes.

The net was beautifully made, with the center opening not more than a quarter of an inch in size. I held it out to her and asked the reason for the hole.

She smiled tolerantly and explained it was where the good dreams came through the net; the bad ones, not knowing how, were tangled in the mesh and when the light struck them in the morning, they died.

I tucked the dream net gently into my pack, tried to show my faith and was rewarded by a smile. That night when I returned, I told my two little boys the legend of the Chippewa and hung the net between their beds. There it stayed for years until the boys grew up and went away. Though they are now far past the age when they can be expected to believe such delightful fantasies, I am sure they have not entirely forgotten and sometimes, when they close their eyes at night, they still may sleep with the calm assurance that their dreams are guarded well. Their years of childhood faith, I believe, left a mark upon them and as proof, the little net was eventually hung over another little bed in far-off Alaska. The net accomplished its purpose as it had for the Chippewa over centuries of time.

What the Indian woman told me spoke of love and tenderness toward children, a trait of character I found in the north and among all Indian tribes wherever I have been. But it was with the Menominees of Wisconsin that I found a translation of an ancient cradle song that to me embodies all of the beauty and poetry of the legend itself.

Ne pa Ko my sleepy head,
In your basswood cradle bed
Down cushioned gently swaying
To the song the winds are playing,
Homeward whippoorwills are winging
Hear them now their sleep song singing—
Sleep, little warrior, sleep.

Ne pa Ko my sleepy head,
The golden bees have gone to bed,
Silver, grey-green dragon flies
Close their luminous great eyes,
Wings of silken moths are still,
Pine birds call you from the hill—
Sleep, little warrior, sleep.

Ne pa Ko, my sleepy head,
Tiger lily's gone to bed
Where the heron tall and stately
Watches for the moon sedately,
In the swamp the marsh hen drowses
Near the muskrat's winter houses—
Sleep, little warrior, sleep.

Winds are whispering in the balsams,
Singing softly in the cedars,
Crooning through the glistening birches—
Sleep, little warrior, sleep.

In the dream net that Indian mother gave me long ago,
I hear this song. Good dreams are there, dreams of birds and butterflies, and the wind over the marshes. When I lie in my sleeping
bag and listen to the soft moaning of the pines, the lap of the
waves, or sounds of night birds in the trees, I often think of it.
Sleep comes swiftly then, for there is a dream net over me wherever I may be. I have thought of it in many places, when bedded
down on the tundras of the far north, in the mountains of the
west, the lowlands of the south, or the highlands of the east. I
have seen it in interlacing branches overhead, in hovering cliffs
and in the snow-clad peaks of mountain ranges, even in the stars
themselves. When I happen to remember, the net is always there,
whether I am sleeping under a tree, against a log, or in some leafy
hollow next to a ledge. On nights when the roar of some rapids we
must run in the morning might have kept me from sleep, it has
been a comfort.

We smile tolerantly at such beliefs but cannot ignore
them, for in all peoples, no matter what the race, there is always

an intuitive sense of a spirit world. To the Indians of earlier days, this spirit world was very real. To them all living things have spirits and one must recognize that they exist and that people must live with them or die. Before any creature is killed, its spirit must be spoken to; before a plant is taken for use, it must be approached with reverence and explanations. When an Indian mother makes medicine for her children, she invokes her own guardian spirit to guide her to the proper plant and for skill in compounding a cure. When she finds what she has been searching for, she invokes the spirit of the plant itself and takes it with gratitude and faith. If death results, it merely means that the great Manitou has called the loved one to the Happy Hunting Grounds for reasons of his own.

Such implicit faith can never be explained through reason or cold scientific analysis. Back of it are a million years of racial experience when men lived as did other creatures, with only one purpose, physical survival: finding enough food, shelter from the elements, and producing young to perpetuate the species. Yet, back somewhere in those dark and misty regions of the past, some mind must have broken from the old pattern of brutishness and showed the first vague intimation of something different, its evidence perhaps nothing more than a gesture of affection or an impulse to share a morsel of food with the dead. Whatever it was, or however it came about, here was a dividing line between primitiveness and a new world of imagination and spiritual awareness, the great leap in the evolution of the race. Granted that creatures other than man show love and feeling not only for their young but for each other, only in man has it progressed to where it is a major force in his development and culture.

All this is a legacy from the dusk where magic once ruled. It is part of the inheritance of longings, hopes, and fears that came to men when they gazed into their fires at night, into the blue distances, or at the stars, and dreamed strange dreams of beauty and peopled the unknown with many forms. Though this was during the childhood of the human race, the wonder and the dreams live on and play a role in all our lives. Here is the creative force in art, music, literature, and religion, the wellspring from which all our progress comes.

Many dreams are those of beauty and, like the Navajo, those close to the earth and not too far removed from the ancient spirit world, can pray with them and say:

Beauty is before me.
Beauty is behind me.
Beauty is below me.
Beauty is above me.
I walk in beauty.

My little dream net spoke of many things to me, of love for children, of tolerance and the intangible qualities that give warmth and meaning to life. When I accepted it, I did so with humility because for me it was a symbol not only of trust and acceptance by my Indian friend, but a hint of the long past and a world of dreams most moderns have forgotten.

Not long ago I slept under a pine tree with my six-year-old grandson, Derek. We had a tiny fire and lay in our sleeping bags, watching the reflection of dying embers against the branches and how it turned them to gold and bronze and copper. I told him many stories of moose and bear and, at the end, the legend of the dream net; he believed, for he was young and still had faith. Above us that night was the ancient net, and the night was full of dreams both good and bad. He could see them, feel them, and when I told him that only the good ones came through, he closed his eyes quietly and went to sleep.

Later when the fire had died and the stars were bright, as they are when one looks up at them through tall dark pines, I wakened him and we lay watching together. His eyes were big and full of wonder as I pointed out the Great Bear, Cassiopeia, and the Pleiades, and the vast expanse of the Milky Way. Once we saw a falling star. It streaked across the heavens from east to west, a glittering cascade of shattered light.

"Papa," he said excitedly, "why did that star fall?"

I tried to explain but his young mind was not ready for talk of disintegrating matter, oxidation, burning in the stratosphere, and the bombardment of the earth with intersteller dust and meteorites. So I simply said:

"Stars fall like the leaves of trees; sometimes they get tired of holding on away up there and just let go, and then they fall."

My explanation satisfied him completely. There was nothing more to say.

"Do you suppose we might find one someday?" he asked finally. And I told him if we looked real hard, we might find one lying in the woods in the morning, and with that his eyes closed again and he was asleep.

I could not help but think as we lay there of what Loren Eiseley said: "Men troubled at last by the things they build, may toss in their sleep and dream bad dreams, or lie awake while the meteors whisper greenly overhead, but nowhere in all space or on a thousand worlds will there be men to share our loneliness."

All this Derek could not understand and it was just as well, but he knew beauty and wonder as he watched the falling star. To him it was only a dream, but nevertheless one with the dreams and hopes of all mankind.

Another night I lay on a lonely point of rock in the Quetico-Superior country. Again the stars were bright and I watched for the meteors I knew would come that season of the year. For a long time there was nothing, only the vast swirling nebulae, then across the heavens went a streak of light. I did not hear it whisper, nor was it green. It flamed for a moment like a torch of sparkling fire and then disappeared. I thought again of Eiseley's words and the consciousness of man's scientific achievements came home to me; the universe now of a size beyond imagining, the earth floating like a grain of dust in the void. Here I lay on a rock, a tiny living speck of sensitive protoplasm and, for a moment, I felt lost and alone with what I knew.

Lying there, the story of the dream net did not seem to matter, nor did any of the other legends, traditions, or beliefs of men in the spirit world. Protons, neutrons, neutrinos, space time, these were the real things—and thinking of what they meant, I was troubled.

But then I remembered a night on a glacialogical survey with the famous geologist Dr. Wallace Atwood. We had found an island with a beautiful outcrop of Saganaga porphyry, something he had wanted to see for a long time. Porphyry had a special meaning for him and he had looked for rare specimens of the rock all over the world.

We sat before our fire that night and talked about the glacial patterns we had seen over the Canadian Shield, but mostly we spoke about porphyry and when he fondled the prize specimen in his hand, his eyes shone with delight.

"Tell me, Dr. Atwood," I said finally, "how is it that at the age of eighty-four, you still get as much pleasure and excitement out of a find like this as though you were a student on your first expedition?"

He looked at the bit of porphyry again, turned it so the light gleamed on its surfaces. It was a perfect piece and I knew it would go back with him when the survey was completed.

"The secret," he said, "is never to lose the power of wonder at the mystery of the universe. If you keep that, you stay young forever. If you lose it, you die."

The power of wonder, imagination—the sense of mystery and belief—that is what I saw in Derek's eyes that night under the pine tree. The same light and faith were in the eyes of the Indian woman of Grand Portage when she said: "The bad dreams get caught in the net, the good ones come through."

I forgot my ponderings and settled down in my bag. Though the simple legend of the dream net was only an infinitesimal bit of the whole fabric of the dream world, a fabric that has brought beauty and meaning to searching minds for untold generations, it was in its way perhaps as important as any other.

Myriads of stars were my net that night, but I no longer felt lonely, for I knew that while man might unravel the puzzled skein of life and solve the riddles of the universe, what really matters is the wonder that makes it all possible. Back of everything is always a net of dreams.

THE GOLDEN AGE
OF THE CANOE

Henry Beston

The "canoe" we know today, though in a general sense the gift of
the Algonquin tribes of the northeast, is more particularly a Chippewa creation, a Chippewa masterpiece. Living in the heart of the
old canoe-birch country, every man of the nation with his eye and
mark on some great tree, the tribe united as no other a special
skill in design with the finest of materials. The lovely curve of bow
and stern remains for us their sign. Every Indian nation of the
birch region had its own native and tenacious image of that bold
symmetry. Some made of it a quasi perpendicular, some put the
depth here, others there: a stranger could be placed by the line of
his canoe as easily as by the cut of his moccasins. To the Chippewas alone was reserved the sense of the curve in its perfection,
in its unique and beautiful rightness. Strong, well-made, capable
of carrying heavy loads yet easy to portage, the Chippewa model
like the covered wagon is a part of the history of the continent.

This was the craft that was to make possible the opening and mapping of something like a fourth of North America. In celebrating the covered wagon we have forgotten a little this episode of the canoe. Enlarged by its Indian builders for the fur trade of the old Canadian northwest, it kept its Chippewa character and strength, making itself a vessel for cargoes and crews without losing one touch of its old beauty of design.

Three principal types were in use in the trade. The great Montreal canoe, or *canot du maître*, intended for use on the larger lakes and more navigable streams, could be anywhere between thirty and forty-five feet long. Such a vessel could carry tons of furs eastward from the posts. Fourteen men made up the crew. The north canoe, or *canot du nord*, was a smaller type; built for use in the wilderness itself, it averaged twenty-five feet in length and carried a crew of eight. Between these familiar models stood an intermediate third, the bastard, or *le bâtard*, which carried a crew of ten. Small canoes such as we know today were also everywhere in use. To judge by many old pictures and sketches, sails were sometimes rigged, being most probably raised up on occasions when it was possible to "sail before the wind" in light airs.

The canoe workshops remained in the birch country and on the Lakes. Once built, the canoes of the trade went in for that liveliness of color which is so good for the soul. It was not for nothing that the later eighteenth century had rediscovered and enlarged the bright possibilities of paint. Gunwales were festooned or spaced in green and white or in red and white, and there was almost invariably a design of some kind—an Indian head, a bear, a sun or a moon with features, clasped hands—painted bow and stern. The paddles of strong red cedar were also painted with stripes and gaieties.

A carry down the beach into the placid water of some cove and the craft was ready for its man. He was at hand. Hardy and enduring as few strains have been in history, unwashed, merry, and famously polite, short of legs and powerful of shoulder, pure French now, and now half-Indian, the canoe had already invented its own human being, the woods their man, the legendary and incomparable voyageur.

Westward beyond the great horizons of Superior, westward beyond the strange, jade-green waters and the tense yet empty wind, westward a thousand and even a long two thousand miles away, the forts and stations of the fur companies stood in the immense solitudes of the forest. From the dying out of the great plains north to the arctic barrens, from the Lakes west to the mountain descents to the Pacific, the wilderness spread wide over a solitude of the continent, a region of lakes and woods, rapids and rushing rivers, bogs and quaking swamps and mountains without a name. Within, there lay hidden a complexity, some valleys and forest floors teeming with life, others strangely with scarcely a sign of anything alive. Till the arrival of the fur trade, nothing that was not a part of nature disturbed a quiet of nature widespread and empty as the sky. Only the nomad Indians of the American north, the Dene, the hardy Chippewa, the Crees, and the western Montagnais were a part of its existence, crossing it with scarce the bending of a branch or a footprint in the leaves.

A great skein of waterways leading west and north out of Lake Superior was the gateway to the mystery. Indians had been the first to use the passage, tying river to lake and lake to river again, and the French had been aware of it since 1731, La Verendrye and his guides having gone in as far as the Lac des Bois.

By the end of the eighteenth century, the fur trade had chosen and made customary a great passage to the woods known as the "Great Trace." It began at Montreal with the waters above Lachine and, entering the Ottawa, ascended that stream of many portages to the Mattawa, a tributary flowing from the west. This in turn led to Lake Nipissing, and from Nipissing, hailing with a cheer a westward-flowing stream, the voyageurs descended to Georgian Bay and the waters of the Lakes. The charming island of Michilimackinac, depot and administrative station of the trade upon the Lakes, next awaited the adventurers: here the "brigades" going into the deeper wilderness separated from those bound to nearer posts. So distant were many stations that it took the best of summer to arrive, and the voyageurs wintered at the forts.

All summer long the pretty island was a scene of bustle and activity. Goods were transshipped, crews sorted out and reassembled, the sick attended to, and canoes repaired. Standing on the heights at night, looking out into the vast darkness above Huron, one could see fires burning all up and down the lower beach, each glow of fire crowded close about with its own company.

For those bound north and west, the next great station was Grand Portage on the western shore of Superior. (It is today a town in Minnesota just below the Canadian frontier.) Here nine miles of rapids on the St. Francis River made necessary a long carry. In the great days of the trade homemade roads had been built at the carry, and a score of wagons and several hundred horses assisted the voyageur crews to move their goods and canoes to the navigable waters. Ahead lay the entering chain of lake and river widenings, the Lac du Bois Blanc, the Lac de la Pluie, the Lac des Bois, and, ultimately, the great lake "Ouinnipique." Beyond lay the unknown, the white streams and the forest-brown, the named and the unnamed, the peaceful and the perilous. At dangerous rapids there were always crosses to be seen against the forest wall, each with its voyageur's cap fading in the sun.

Standing near the greater portages and by the junctions of streams, the forts of the trade awaited their first arriving hail.

Each had its chief, or *bourgeois*, usually a Scot, each its tallymen, clerks, and accountants, each its population forever changing and mingling, of trappers and hunters, half-breed children and Indian wives, voyageurs, scouts, and forest adventurers. The Indians and half-breeds were usually the trappers, taking the animals in winter when the pelts were at their best. In the spring, bands would arrive with their catch, the furs hanging behind them from their shoulders. Such a population lived as it could. Now buffalo meat and deer went into the pot, now flour, grease, wild rice, and a bear's haunch all cooked together into some hearty Indian mess. (An Indian stew can last for years, seemingly recreating itself miraculously from the bottom of the pot.) With nothing but the forest about it for a thousand miles the fort lived its vigorous life of direct contacts, slept in its blankets and buckskins, drank its rum, smoked its tobacco, wrestled out its male rivalries, listened to its interminable Indian legends, and married "according to the custom of the country." Parentage could be vague. "*Que voulez-vous?* What d'ye want, laddie?" said one Scot trader to an Indian boy. *"Monsieur,"* replied the youngster with gravity, *"vous êtes mon père."*

French was the common tongue. Wild and outlandish as such a life must have been, it is clear that it did not become barbarous. The natural good manners and sociability of the French Canadian kept it all a remarkably good-tempered adventure. It was with a gesture of politeness that one was offered a little more of the bear.

With portages to make and currents to battle, with loads of supplies to carry in and heavy furs to carry out, with the wilderness for a country and elemental danger ever near, the life of a voyageur was no adventure for the weak. In good weather and when not fighting a wind or a stream, a crew could paddle fantastic distances. Between earliest dawn and summer's dark, canoes often managed sixty, seventy, or even eighty miles. One observer speaks of about forty strokes of the paddle to the minute—a brisk rhythm and speed. Two meals a day were eaten and after a hard carry, a third. They ate everything. Pemmican, fish, birds' eggs, and almost any kind of bird, hawks among them, squirrels, porcupines, dough cakes, and grease dumplings—all these were downed with relish by the evening fire.

The contemporaries of the voyageurs who accompanied them on their expeditions above all remembered the singing of the crews. Mile after mile they sang, singing together with the thrust of their swift strokes, the gay, choral sound echoing back upon them from the enclosing walls of the forest or floating off across the stillness of lakes into the north and the unknown. It was ever a cheerful sound, a sound of labor and the human spirit, a music of the body's good will and the heart's content. Old ballads and songs of France made up the substance of the singing, most of them unchanged in verse or tune, though now and then a wind from the spruces had blown across a song, making it more Canadian in its language and mood. It is a man's world that is here reflected; its concerns are going courting, the formal elegancies of wooing, the pains of youth and broken hearts, and noble and ceremonious farewells. Nothing can exceed their decorum. To this pleasant and old-fashioned treasury the voyageur came in time to add new songs of his own but the old songs remained his favorites. If they were not Canadian in the beginning, he made them Canadian by adding himself.

So the cavaliers bow, sweeping off their seventeenth-century hats to tunes made for harpsichords, the lover laments, and the soldier returns from the wars. And all the while the forest passes by, the white water rolls over the rock, the sides of the canoe scrape with a rasp through the pitcher plants, and the paddles dip and thrust and rise gleaming together in the sun.

No adventure of the Canadian past so stirred the heart as the departure of the voyageurs from their depot at Lachine. One came upon them in the busy spring, some camping by the river in the open fields, Montreal and its church bells behind them to the east, and before them the afternoon sun and the adventure of the west. For days before the embarkation wagons had been arriving with their loads, rolling through the farming villages and deepening the ruts and puddles with their weight of trade goods, provisions, and supplies. In and out of the offices and wharves, busy at a hundred tasks, yet always finding a moment to toss back a jest, swarmed the adventurers, a whole French-Canadian countryside of Gaspards, Aurèles, Onesimes, and Hippolytes. There was much to be done. Here, on the beach, men crouched by a canoe, making some last repair; here clerks scrambled over boxes and bags checking and rechecking the trading goods, the trinkets, beads, axes, knives, awls, blankets, and bolts of bright-red English flannel; here an official studied the enlistment papers of some new engageé. At a counter to one side, a crowd selected the shirts, trousers, handkerchiefs, and blankets due them from the company, Iroquois Indians from Caughnawaga, famous paddlemen, reaching in and seizing with the rest. Late in the afternoon, those who were quiet over a pipe could hear the eternal murmur of the miles of rapids, and the floating, clanging summons of the Angelus.

The moment of departure waited upon weather and the wind. To prevent a last and too-thirsty festival of farewell, efforts were sometimes made to conceal the probable day, but men concerned have sixth sense in these matters, and the world was apt to share the secret, and all Montreal, finding the morning fair, came to say good-bye. Ladies with escorts watched from the shores,

British officers, mounted on English horseflesh, rode to good places in the fields, British soldiers even, their flaxen hair and blue Sussex eyes a new note in the throng, strolled in pairs among the Indians. Citizens and citizenesses, wives and children, parents and kin, company directors and curés—all these were at hand to see the start. It was early May, and the Montreal country had left winter behind and was taking courage in the spring; on far shores and near, under the cool wind, appeared the green.

In and out of the press, heroes of the occasion, moved the voyageurs. Old hands and new, it was their day. Even the young Scot clerks who were to go as passengers to the forts shared the importance and the glory. Custom demanding that the beginning and end of a journey should be carried off in style, every voyageur was dressed in the best he had. A woolen tunic or long shirt worn outside and belted about with a bright, home woven sash—the charming, old-fashioned ceinture *fléchée*—Indian leggings or even homespun trousers, a red knitted cap, and heavy-duty Indian moccasins—this was the costume. A beaded Indian pouch worn at the waist, Iroquois or Chippewa work, was a particular *sine qua non*: indeed, all veterans were engayed with Indian finery. Voyageurs belonging to the governor or chief factor's brigade had feathers in their caps. Often a small British flag was flown from each canoe. The fleet sailed by "brigades," by groups under one command, and these kept together, maneuvering with careful paddles in the current falling to the Lachine. Are all afloat, all loaded, all officers and passengers in their seats? Then go! Church bells rang, guns were fired, and on the broad river paddles dipped and thrust forward in a first strong, beautiful and rhythmic swing. At the same moment the river covered itself with singing. The fleet beginning to open, the brigades sorting out, one could see nothing but canoes for miles, hundreds upon hundreds of the laden craft all striking as one into the purplish-brown waters of the Ottawa.

At the northwestern corner of Montreal Island stood a church of Ste. Anne, patroness of sailors and of voyageurs. Here the brigades made a first halt and landing, the paddlemen and bowsmen, the steersmen, clerks, and passengers all trooping up from the beach to pray for a safe voyage and a safe return. It was the custom to make some small offering, and the Scot Presbyterians, it is said, made theirs in propriety with the rest. Soon they were all of them on the river again, the church hidden by some turn of the stream, some brigades falling into their measure and stroke, some out of high spirits leaping ahead with a song. *"En roulant, ma boule, roulant,"* and out of sight they go. Thrust by thrust, by quiet waters and by furious streams, through the summer plague of the stinging flies and the blessed coming of the early cold, the paddles will swing across the half of a continent, making their way into the forest, into the land of Keewaytin, the northwest wind, the ancient land where nothing has changed since the beginning of the world.

GLIMPSES OF THE PAST

Grace Lee Nute

The French Regime

Hardly were the Pilgrim fathers acquainted with their rocky fringe of continent when French explorers reached the very heart of North America. By 1660 both shores of Lake Superior had been visited and men had gone beyond—how far we do not know. On seventeenth-century maps appeared a "Groseilliers River" on the north shore of Lake Superior. Whether or not this was the fascinating river now known as the Gooseberry is uncertain, but the name may well designate a stream visited by the Sieur des Groseilliers in the spring of 1660. He seems to have explored at least a part of the north shore that spring in company with his young brother-in-law, Pierre Esprit Radisson, and they may have used one of the several Indian portage routes from Lake Superior to the Rainy Lake–Winnipeg River canoe route to the West. Some historians are inclined to the view that the Groseilliers River of the early maps was the Pigeon River of today and that the two brothers-in-law actually knew and used the famous Grand Portage at the mouth of Pigeon River. Even if they did not venture inland toward Rainy Lake, it was only a short time before a Frenchman *did* explore that ancient canoe route along Minnesota's present northern boundary.

The first Frenchman actually known to have ventured from Lake Superior over the canoe route toward Rainy Lake was a resident of Three Rivers in the province of Quebec, which was a prolific nursery of explorers and voyageurs. His name was Jacques de Noyon and early documents refer to him as a "voyageur," a term that soon came to have a special meaning. In ordinary French it means merely "traveler," but in North America it meant a canoeman in the fur trade. There are several references to de Noyon in French records, but one in particular, written in 1716, refers to him as having wintered "about twenty-eight years ago" in the Rainy Lake country on the "Ouchichiq River." This was probably an early name for Rainy River, since it is mentioned as leading to "the Lake of the Assiniboin [presumably Lake Winnipeg] and from there to the Western Sea." A document of 1717 states that "some voyageurs have already been clear to the Lake of the Assiniboin."

After the 1680s the French records are silent about the region for thirty years or so, though it is almost certain that other white men penetrated the area west of Grand Portage during those years. Then, in the years after 1731, the boundary waters were the usual thoroughfare to a West that became better and better known as the years rolled by. Explorers and fur traders found routes, built forts, established practices and customs, and gave names to physical features. It was in 1731 that Pierre Gaultier, whose title was Sieur de la Vérendrye, began his explorations beyond Grand Portage; and it was about 1760 that the last French post was abandoned in the *pays d'en haut*—the "upper country"—as the traders called this region and other parts of the West of that day. So the French regime may be said to have lasted just a century in this borderland.

The British Regime

With 1760 came the conquest of Canada by the British. France lost its fine, great colony in 1763. English and British colonial explorers and traders now became common on the border waters. Though on paper a large part of Minnesota passed to Spain in 1763, and though the Americans between 1776 and 1783 fought and won a war with Great Britain, this border area remained practically British till the close of the War of 1812. Kings came and went, governments rose and fell, wars were fought, and boundary lines were placed at will, but the border country cared little. Its life went on as before, full of activity, danger, adventure, the struggle for existence, the round of ordinary daily life in a region that was virtually a law unto itself.

The period after the Revolutionary War witnessed the interplay of three great trading companies in the region: the Hudson's Bay Company, the North West Company, and the American Fur Company. It was also the heyday of the voyageur, that dauntless mariner of the western waters. Over the lakes and streams of the north woods floated his light bark canoe propelled by red, flashing blades to the tune of his Loire Valley *chansons*.

About the year 1768 John Askin cleared the site of Grand Portage, at the Lake Superior end of the border-lakes canoe route, to make ready for a post there. From that time till shortly after 1800 Grand Portage was the great inland depot of the North West Company's fur trade. Other companies and individuals also made it the center of their operations. Every July a great gathering of traders and voyageurs was held there, to which men came from far out on the Great Plains, from the Oregon country, and from bleak tundras beneath the Arctic Circle. A large part of the trade of a vast hinterland, embracing the center of the continent and many of its fringes, passed through this narrow channel on its way to the great fur marts of Europe and Asia.

Another post was established at Rainy Lake. To it the men from Athabaska, hundreds of miles to the northwest, could travel and return in one season; but if they went on to Grand Portage, they might find western lakes and rivers blocked with ice on the return trip. So an Athabaska House was a prominent feature of the Rainy Lake post. A special set of canoemen from Grand Portage met the Athabaska men there, received their packs, and gave them their annual supplies in return.

Most of the border-lake posts were dependent on one of these two forts. Even the Lake of the Woods posts were, as a rule, only wintering houses from the Rainy Lake fort. There were few forts, moreover, that did not lie along the boundary waters. Red Lake, Leech Lake, and Vermilion Lake had posts; the country between them and the border waters had few, if any.

The North West Company had things pretty much its own way on the border till 1793, when the Hudson's Bay Company, long quiescent along the shore of Hudson Bay till stung into action by its rival, began to establish posts at Rainy Lake and on Rainy River. Competition flared, particularly from 1793 to 1798 and again from 1818 to 1821. Between 1798 and 1804 the North West Company's chief rival here was an offshoot of itself, generally called the X Y Company. After the War of 1812, and particularly after 1823, Americans of John Jacob Astor's American Fur Company were the principal competitors, with posts at Grand Portage, Grand Marais, Vermilion Lake, Moose Lake, Basswood Lake, Rainy Lake, Rainy River, Warroad, Roseau Lake, and Lake of the Woods. So keen did the competition become that finally in 1833 the Hudson's Bay Company, which had absorbed the North West Company in 1821, bought off the Yankee traders with an annual payment of three hundred pounds sterling. The Hudson's Bay Company posts were practically the only ones on the border lakes between 1833 and 1847. This company had few posts south of the border at any time, except in the valley of the Red River of the North and on one or two of its affluents. Even these were short-lived. When Minnesotans speak of old Hudson's Bay Company posts on Minnesota soil, they almost invariably have in mind forts of the North West Company or of the American Fur Company.

The American Regime

The Yankees who followed close on the heels of the British traders and explorers witnessed many boundary disputes between the

United States and Great Britain in the twenties, thirties, and forties of the last century, while the first infant settlements were burgeoning into life in the region that in 1849 was to be named Minnesota. Attempts to settle these disputes were made again and again, by treaties, conventions, boundary surveys and commissions, arbitration by foreign princes, and other expedients, till in 1842 the Webster-Ashburton Treaty laid the basis for a lasting peace, the settlement of the boundary dispute, and many years of close friendship between the United States and Canada.

Geologists became the next explorers of the region, particularly between 1848 and 1880. A false gold boom, begun by one of them, led to a temporary opening of this frontier to settlement at the close of the Civil War; but it was really the discovery of iron ore that brought permanent settlers to the region in the eighties and early nineties. From the villages of the Vermilion and Mesabi ranges, which were founded at the time of the mining booms, people from Finland and southeastern Europe gradually slipped out into the wilderness, cleared little farms, and began a very independent sort of existence. Their culture has become a unique part of Minnesota's life.

The logging boom then came upon the region early in the twentieth century, when lumberjacks by thousands cut the great pines in winter and often labored as harvest hands in the Red River Valley and on Dakota prairies, or on whalebacks and other ore boats on the Great Lakes, in summer. These red-shirted, stag-trousered, calked-booted men were close rivals of the voyageurs in romance and hard, dangerous living.

The monarchs of the forest fell. Americans began to see the loss they had sustained in the deforestation of most of the north country. Just as the First World War broke in Europe, a conservation program was initiated in Minnesota to preserve its remaining forests, lakes, streams, wild animals, flowers, birds, and wilderness.

ADVENTURES IN SOLITUDE

Calvin Rutstrum

To follow a wilderness trail over a route of no special itinerary, un-mindful of tomorrow's prospects, to be wholly unapprehensive of life's daily fortunes or misfortunes for a while, this seemed to offer the brightest alternative from my recent military confinement and the disrupting effect of a world war. This time, I would leave the canoe behind and travel with a light camp outfit and whatever food I could carry in a shoulder pack, otherwise live as best I could off the country.

A fish company steamer, the *America,* was plying about two hundred miles of Lake Superior's North Shore from Duluth to and around Isle Royale. A dirt road followed Superior's North Shore but was mud-mired and offered no commercial transporta-tion. While the *America* carried both passengers and freight, the collecting of fish from Scandinavian commercial fishermen who were settled along the shore provided the boat's chief revenue. In many stopping places there were no docks. The steamer, unanchored, would shut off its engine and lay adrift on the swells, while a lone fisherman would row out to the ship in a skiff with a load of fish and return with empty herring kegs and trout boxes. Numerous gulls would arrive from nowhere, bobbing on the swells near the ship, expecting to be fed by the passengers.

My own destination was Hovland, a small fishing settlement about a hundred and thirty miles up the lake from Duluth. The steamer was scheduled to arrive at Hovland on an average of twelve hours from the time we embarked, but two hours or so from port a storm began to build up. At first there were only high, heaving swells, suggesting that the weather had become rough on some distant part of the lake. My navy-acquired sea legs proved to be of some advantage. I learned that most of the passengers had become seasick from the alternate rising and sinking movement of the ship on the swells, and had taken to their cabins. In about five hours the storm bore in on us in earnest, waves of spectacular height breaking with a thunderous boom as they burst with explosive, sky-leaping, whitewater displays against the high rock escarpments. The ship pitched wildly, its bow seeming destined to submerge one moment as breakers streamed over it, the next moment its bow was reaching for the sky as though about to take off in flight. One wondered if the ship, creaking in protest, had a backbone stiff enough to withstand the strain. Such strain, I was told, sometimes broke steampipes, leaving the ship without power to keep it from shore and headed into the sea. The thought of a ship without power being smashed against some rocky palisade was not exactly comforting.

The safest place from the heavy wash of the sea itself was below deck, but I managed to ingratiate myself with the ship's Indian skipper, and rode out the storm with him, high up in the wheelhouse—a box seat to a spectacular natural whitewater production. When the skipper learned that I had just been discharged from the Navy and that I was headed into back country with a shoulder pack, we found a warm comraderie that seemed to emanate both from his seafaring and Indian, wilderness background.

He told me that Hovland would have to be bypassed, the sea being too heavy for a safe docking there at the boulder and dirt-filled log crib. I would be carried on around Isle Royale to be let off at Hovland on the return trip. This pleased me because I had not seen Isle Royale before—a strip of land about forty-five miles long, lying fifteen miles out from Canada's North Shore of Lake Superior, although the island itself is actually a part of Michigan, which borders on the south shore of the lake. Besides, I was hell-bent for nowhere in particular, as pleased to be diverted by the elements as by any other deviating circumstance. Moreover, I would from now on until Hovland was reached, be the free guest of the fish company and their sumptuous ship's cuisine. When the ship was brought in to the lee of Isle Royale, it was anchored there for the duration of the storm. One knew only by the smooth, high-riding swells, and the wind's incessant roar above the trees, that a storm raged elsewhere. The ship's position offered no relief in its uneasy anchorage for the seasick passengers. Except for a member or two of the crew, I dined alone.

Passengers were largely in bed, although a poker game was going on in the ship's hold on a cargo of sacked oats, flour, and meal. A fight broke out when some Finlanders discovered that they had been taken in by two itinerant cardsharks. As knives began to flash from Finnish sheaths, the cardsharks fell back in a frightened huddle. Once the ship reached a dock on the main shore the following day, the cardsharks took off, penniless, a bit bloody about the face from a beating. They were last seen heading toward a dirt road near some fishing huts. Apparently, they were glad to be ashore, considering it was a lesser consequence than possibly being turned over to the sheriff.

I was to be landed at the Hovland crib dock about midnight, the sea now growing fairly calm, though the sky was still heavily overcast. The skipper would sound the ship's horn at regular intervals, then wait for the echo to come back from the forest wall. By long experience, he was thus able to gauge from the lapse of time between the horn blast and the return echo, how far out from shore the ship was. Nothing could be seen in the darkness. Later a mere firefly speck of light began to appear in the distance. It proved to be a single kerosene lantern that identified the landing spot for Hovland. I watched the tiny glow steadily as it grew in size and importance to me.

A lone fisherman met the *America* at the dock. Rapid unloading of empty herring kegs and knocked-down trout boxes was followed by a loading of salted herring in kegs and iced trout in boxes. The *America,* late on its schedule because of the storm, was being hurried on its way. Its lights soon disappeared around a headland, leaving the dock in darkness except for the kerosene lantern that now seemed to grow brighter by the moment and illuminated the whole dock area as one's pupils dilated and eyes adjusted to the darkness.

I had expected to pitch an overnight camp nearby, but the fisherman, in a heavily Swedish-accented voice, assured me that this was not necessary, that a bed would be found somewhere, once he had the kegs and boxes stowed away. He also commented on the possibility of rain. I set my pack aside and proceeded to help him with the stowing job.

Well up on the shore slope, I noticed the lighted windows of a house. Our stowing job done, the fisherman headed for the house and asked me to accompany him. Through the open door I could see a table set for coffee on a brightly checkered tablecloth.

In Swedish, he addressed his wife. "Mama," he said, "we have company. Maybe he would like a cup of coffee too." He looked at me somewhat amused and remarked, "I guess you don't understand that kind of language." When in the best Swedish I could muster, I told him that I had understood, whatever barriers of strangeness existed quickly vanished.

The conifer forest at night and the freshness emanating from the cold spring water of Lake Superior gave an inspiring seaside fragrance, but when I caught the aroma of coffee and saw homemade *smor bakalser* ("butter rolls") gracing the table, I would have been hard put to make a decision as to which was most pleasing to my olfactory sense. I apologized for my late-hour intrusion. However, I was soon assured that I would have a good bed and that company did not come too often in the woods, "so one could not be too particular about what time of day visitors arrived, could one?"

It seemed that all the pieces of a delightful pattern fitted into my arrival until I mentioned that I was heading north on the trail to Pine Lake and the boundary. My host looked at his wife for a moment, then turning to me, said, "Don't go there." He told me that a feud had been going on between a trapper who lived at Pine

Lake near the United States–Canada boundary, and some people nearer Lake Superior. The trapper, he explained, had been intimidated somehow and had apparently threatened to shoot anybody who came into his general area. I would likely get shot, my host feared, if I continued on up the Old Trail to Tom Lake, the halfway point from Lake Superior to the boundary.

Since I was wholly alien from the local population, I saw no reason why anyone should involve me in the feud. After a good night's sleep and breakfast, I bought supplies at the local post and headed north on foot with a well-loaded shoulder pack to the first stage of my hiking trip, Tom Lake, eleven miles up the Superior watershed.

About halfway up on the north side of Tom Lake, I found a small abandoned cabin with a split-log floor, furnished with a rusty sheet-metal stove, a bunk built of poles and balsam boughs, a table made from poles, and two wooden boxes for seats. The cabin had a narrow swinging door made from hand-split, half-round logs, but no window. Once the door was closed, all was darkness, except for light leakages where the log chinking had fallen out.

I started a fire in the stove and prepared a meal, the door left open for light. A rather heavy rain fell during the night, but oddly the cabin did not leak. I was sure that if it did, the trickle would, by the law of averages, be on the bed. Therefore, I was prepared to drape my tent, if necessary, over the bunk like a canopy. A few creatures ran in and out of the open door during the night, including a porcupine that chewed disturbingly for a while on the table, where, no doubt, he found the taste of salt. Little, however, was apt to disturb my sound sleep for long, after carrying a pack twelve miles over a wilderness trail.

The next day dawned bright under a blue sky with a few cirrus clouds. I rose, got a fire going in the stove at once for coffee, bacon, and pancakes, looking forward to whatever events lay in prospect. A red squirrel ran in through the door in short spurts and pauses, apparently having been fed by a former occupant. The squirrel ran up the walls, over the bunk, and finally landed on the table where he took bits of food from my hand.

While I was frying pancakes, the cabin suddenly went dark. At first I suspected that a black bear might have smelled the brown sugar melting for the pancake syrup and decided to look in. My next thought was that possibly the trapper against whom I had been warned had come to pay me a visit—not, I hoped, with intent to carry out my demise. Since I seriously considered that it might be a big black bear, I did not move hastily from my position facing the stove, lest I should suddenly find myself competing with him against odds for the pancake syrup.

As I turned slowly, I saw a rather swarthy-looking individual standing in the doorway, a Winchester rifle in his hand that pointed toward the floor, not, I became comfortably aware, toward me. I managed a "Good Morning!" as cheerfully as I thought advisable, without appearing patronizing, but received no return greeting.

The psychology under such circumstances, I assumed, was to go about one's business as though all's well with the world. Since the man continued to stand there, I poured some coffee in a tin cereal bowl and also in a split-handled tin cup.

"Have a cup of coffee," I offered.

No answer.

"You must be the trapper from Pine Lake," I said.

"How do you know me?" was his first response and not too pleasantly voiced.

"I just landed at Hovland on the Steamer *America*. When I told people that I was coming through here on my way to Canada, they told me that I had better not go up into this country if I didn't want to get shot by the Pine Lake trapper. Since I have never known or had any trouble with a Pine Lake trapper, I paid no attention to it. Have a cup of coffee."

He set his rifle in a corner near the door closest to himself and sat down on a box at the improvised table. I shared my bacon and pancakes, making more of everything, the stove being within arm's reach of the table. (Whoever had arranged the few simple items in this cabin had a talent for convenience.) No words were spoken while we drank coffee and ate the pancakes and bacon. Without appearing to do so, I averted any possibility of getting onto the subject of his trouble. I asked him about the country, the best route into Canada, and who owned the cabin I was occupying. The sympathetic nature of his answers convinced me that here was no man to fear unless you happened to be his sworn enemy.

By the time a second batch of coffee, bacon, and pancakes was finished, he had voluntarily filled me in on the alleged feud. Two army deserters, he said, had accosted him and his wife in his cabin on Pine Lake, demanding a food supply and money under threat of death. While in a careless moment of both watching their prisoners and searching the cabin, the deserters were distracted, the trapper and his wife made a run for it through the door, snatched their snowshoes from a snowbank, and over a devious route behind the cabin managed to elude their pursuers, who now had no snowshoes on which to follow. Later, returning to their cabin, they found smoke rising from the chimney. A wait in ambush showed that the deserters were still there. A shooting ensued in which one of the deserters was badly wounded and soon afterward committed suicide with a .45-caliber revolver.

The trapper was tried in a rather involved case where public duty and extenuating circumstances became the issues. And although he was subsequently freed, a long chain of perturbing events resulting from implications of the trial somehow made the trapper become feared by the local populace. The strange aspect of this whole episode was that in the end the trapper lived down his difficulty and eventually became a county commissioner. I had met the most feared man in that northern wilderness and found him to be not only a delightful individual, but he later became a close friend.

His parting words as he left me at the cabin on Tom Lake were a concern for my own needs. As he left, he inquired, "You not have meat?" I remained at the cabin several days. One evening on returning from an all-day trip, I found a cloth flour sack hanging from the ceiling. The sack contained about thirty pounds of moose meat, well cooked and ready to eat. Like the native who more conveniently moves to his kill rather than moving his kill to his camp, I remained at the cabin until most of the meat was consumed, holding chiefly to a diet of meat and tea, supplemented with blueberries, which grew in profusion nearby.

WHITHER AWAY

E. F. ("Pipesmoke") Cary

The Namakan River running from Lac la Croix to Namakan Lake is not a well-traveled route—at least it did not have that appearance when we passed through that way. There were few well-established camping places. We did find an acceptable spot below Snake Falls for our first night on the river where a run of small walleyes dimpled the stream surface, apparently feasting on a timely insect hatch, and we had a little fun teasing them with our lures just before dark. On our second day we were not so fortunate and in spite of a mean headwind that bothered us on the more open stretches, we decided to keep going until we reached High Falls, where we felt sure we could find suitable quarters.

Our judgment was excellent, but our luck seemed to run out—as we came in sight of the headland above the roar of the falls, weary with the long day's paddle, we could see in the distance two white tents that were set on an open green, and on closer approach, we noticed, down by the water's edge, a spot of color and movement that suggested human activity. We rather expected an Indian encampment, but on coming in for the landing, discovered, in unexpected surprise, two very pleasant and domestic-appearing ladies attired in house dresses and ample aprons, who were occupied in a very prosaic household task . . . they were peeling potatoes and preparing vegetables for the evening meal.

We got by these good ladies after exchanging a few words of friendly greeting, but did not get by Uncle Neal Berger, who proved to be the genial robust gentleman whom we came upon as he was cleaning fish at the far end of the portage. When we inquired about possible camping places farther down the stream, he wanted to know what we found wrong with the one where we were at and he insisted that we go back and share the accommodations of that green and grassy meadow where his party had already staked out their prior claim. We needed very little urging . . . although burdened with our packs, we have never retraced our steps over any portage with more relieved alacrity nor right goodwill toward our fellow human beings.

We later learned that the Bergers, as a young married couple, had come into the Namakan Lake country by canoe in the year 1892. They had built their cabin on the east end of the lake on the Canadian shore and had settled down to live their lives and to rear a family in this border land of lakes and forest. Each of their boys and each of their girls had gone out in turn into the states for schooling and a higher education, but all had come back to make their permanent home by popular choice in the country of their birth and early environment. As time went on, some of the Berger kinfolk had built vacation cabins nearby so that in the summer months there is now quite a colony established at or on the islands near the mouth of the Namakan River. To all of these people Neal Berger was "Uncle Neal," and we were very happy to be included as guest members of this friendly family group.

After "Uncle Neal" had shoo'd us back over the portage at High Falls and while we were putting up our tents and frying crisp walleye steaks for the unusually late evening meal, we began to sort out the individual members of this party that had made us welcome. Mrs. Berger was one of the two good ladies with the very domestic aprons whom we had first met at the landing. Her companion proved to be the principal of a public school in our home town . . . she owned a very charming little cabin on one of the pine-clad islands in the lake that we were yet to explore. This excursion up the river had been organized as a parting bit of entertainment for two of her visiting friends—also school marms. Along toward dark the fishermen came in. The big, strapping, rough-and-tough-looking fellow with the exceedingly black whiskers was introduced as another cabin owner—it never pays to rely too much on circumstantial evidence when you are up in the woods—this rather careless and unkempt individual turned out to be a prominent lawyer, very keen, well polished and able when practicing his profession, and a most hospitable host. His companion was a young chap, also related to the Bergers, who had come alone all the way from Fort Francis, traveling east by canoe through the tremendous length of Rainy Lake to reach and cross the almost equally wide stretches of Lake Namakan. The name of this young fellow was "Rusty" Myers and we will bump into him again on later portages.

As we were enjoying our late community dinner in the dusky and uncertain light that had come upon us, another canoe party appeared . . . suddenly and in silence an Indian boy came over the crest of the portage hill carrying a canoe and an Indian girl followed closely, holding a small blanketed bundle in one hand and a .22 rifle in the other. They seemed rather embarrassed to be passing before such a large and unexpected audience and Uncle Neal went down to talk with them for a few moments before they shoved off. We will always remember the short, quick, powerful strokes with which this young Indian couple finally got under way . . . so rapidly did they point and propel their light craft away from shore and out into the protecting darkness, that they faded from our sight in but a very few moments.

Uncle Neal reported back that these two young people had just been married and had come up river on their honeymoon to get a deer. Their equipment consisted of one canoe, one .22 rifle, one blanket, and one beef heart that was stowed away in the blanket for food. We wonder how many white folks would have had the courage to set out on such an adventure with such primitive supplies, especially in the chill of a late September night? Perhaps they, too, had expected to camp at High Falls.

That evening was most memorable. When two groups of congenial people with similar interests are thrown together for the first time, especially when bound so utterly together way out in a wide and mysterious wilderness, old tales can be told that are new to some, and, because that is so, all others in the know derive an equal amount of enjoyment as the familiar themes unfold. "Uncle Neal" was the champion storyteller and we all stayed up much later than usual—until darn near the midnight hour in fact—as long as he would spin his yarns. He had an almost inexhaustible supply of stories about his hunting and fishing, about the Namakan Indians that lived nearby, and about the city sports that had come to his door. One incredulous tale we will try to record.

Uncle Neal happened to be down at the shore near the end of his wooded point, which extends into Namakan Lake, when he was hailed by a couple of strangers that were cruising by. . . . "I say, old codger, can you direct us to a good place where we can spend the night?" If the rather undiplomatic gentleman had only not said "old codger," he would have received a very cordial and spontaneous invitation to land at any one of several nearby camp locations, *but he did make that fatal slip.* Uncle Neal pondered a moment because the salutation had rubbed him the wrong way and then replied, "Well, there are a few spots around, but I would not do any camping out if I were you—the wolves are pretty bad right now—they say there's a big pack hereabouts and I've seen a few myself!" This casual, but electrifying statement brought the newcomers to attention, bug-eyed and concerned. . . . "Oh! Is that so? What can we do? Where can we go? Can we sleep in the house?" "No, we're all filled up right now with the family home and I don't know just what to suggest . . . you can take a chance, of course, but it would be rather dangerous to remain out all night!"

We cannot compete with Uncle Neal's story-telling art as he developed this little tale . . . how the strangers kept pleading for shelter and how Uncle Neal kept putting them off . . . at the same time playing on their imaginations with what he claimed to have heard about these local wolfish apparitions—their shadowy, slinking shapes—their starry eyes that glittered in the darkness . . . even Uncle Neal's listeners around the camp fire began imagining things that might be lurking in their own dark neighborhood. Finally Uncle Neal agreed that the now thoroughly apprehensive men might sleep on his front porch, although even so, he was not fully sure that the screen would be sufficient protection. Whereupon, the would-be campers happily gathered up their equipment and the trio started toward the cabin. As they were proceeding up the wooded path, Uncle Neal kept the subject of wolves alive in their conversation and as luck would have it, a great white owl, near at hand, let loose with its peculiar piercing scream! "Good Gosh!" cried Uncle Neal, "There's one of the critters now!" and began lumbering up the trail. He had only taken a few steps when those two city sports whizzed by, frantically spilling their equipment for greater speed . . . and all this had had to happen because one of the strangers had referred to Uncle Neal as an "old codger." It pays to mind your manners wherever you may be—or go.

We fished the eddy below the falls for several hours the next morning and then dropped downstream to Namakan Lake, where we camped about a half a mile from the Berger homestead. Our newfound friends followed later in the afternoon and we heard them go by our camp some time after dark. It was necessary for us to stay three days in the camp that we had made on Tar Point, as the wind from the west blew steady and strong and we dared not venture out on that wide expanse of water that rolled in great waves between us and the islands that were barely visible on the distant horizon. Finally, on the fourth day, when our time had run out, Uncle Neal unlimbered his big boat, which he used for his commercial fishing—it seemed like an ocean liner when compared with our own canoes—and he ran us down the twenty miles to our starting point on Crane Lake.

We have never been wind-bound, before or since, under such pleasant circumstances. The first day happened to be Uncle Neal's birthday and we were all invited to a great big family dinner . . . on the second day we were the guests of our newfound lawyer friend . . . and on the third day we taxied over to the cozy little cabin on the island. That dinner at the Bergers was a memorable occasion . . . one expects the little niceties of social entertaining at home but to come upon them in a frontier cabin after the rough camp life we had just led, was so much more impressive.

We quickly and naturally put on our very best manners. There was one great table, sufficiently large to accommodate the entire party, dressy in its spotless white linens and polished silver and bouquets of fresh-cut flowers . . . we sat in real chairs . . . and we ate of such unusual and tasty dishes — the delicious white meat of sturgeon — preserved and tender pieces of moose meat — a bountiful selection of vegetables from the Berger garden — a profusion of jams and jellies and relishes . . . and then Uncle Neal passed around real cigars!

We have always been a great hand for bringing back a souvenir or a trophy from each of our trips in canoe country, and we have had to take a considerable amount of ribbing because of the weight or bulk of the articles we have been willing to carry as extra burdens in our personal pack. Our collection includes a wooden yoke equipped with home-forged chains and hooks with which some northern trapper once fetched his water in pails . . . a pair of bookends selected from loose rock formations at Slate Island on the Manitou . . . two weathered logging chains found on separate occasions and localities . . . an elongated rock that has the appearance of a stick of wood — we picked it up three times to put it on the fire, the fourth time we placed it in our pack instead . . . another sliver of rock to match with which we beat time on a tin can while Martin showed us his version of the Frog Creek Charleston . . . a rolling pin made from a smooth beaver cutting — and a rusted beaver trap . . . little balsam trees, now grown big, brought back in emptied provision sacks . . . the tent poles we used on our last night on Saganaga . . . and in addition, we have what is familiarly known as a "whiffle tree."

That whiffle tree had, unfortunately, come into our possession in the first few days of our vacation trip in the year we were wind-bound on the Namakan. It made its appeal as an appropriate trophy because it would, like the previously found wooden yoke, make another interesting support for drop lights in our summer cabin. It was discovered on one of the Loon Lake portages and we carried it on all our explorations in Lac la Croix. We were able to park it for a few days while we made the three-day circuit around Irving Island but it became an awkward burden once again, when we came back and cruised north and west on the Namakan River. Somewhere between High Falls and our camp on

Namakan Lake, the whiffle tree disappeared and we were consequently most greatly grieved because we had carried it so far and it was such an interesting trophy. On our last morning as guests of the Bergers and after eating a most bountiful breakfast of old-fashioned sourdough pancakes, Uncle Neal let his good nature get the best of him . . . with appropriate ceremony he brought forth and returned to us that lamented and missing object of our affection. He had come upon it at the down-river end of the Hay Rapids portage, where we had stopped for lunch, and because it had appealed to him as being a very practical and useful piece of equipment, the whiffle tree had subsequently been able to continue its interrupted journey down to Namakan Lake.

It has indeed been a pleasure to have known, and to remember our visit with, the Berger family. One of our metropolitan papers once published a story about Mrs. Berger. . . . It seems she had gone all alone to some distant place traveling by dog team in the dead of winter, to attend a meeting of her ladies' aid. This little incident appealed to city people as an interesting bit of news but Mrs. Berger will tell you it was really nothing out of the ordinary . . . someone in the family, usually the youngest daughter, Elizabeth—made the forty-mile round-trip run over the ice to the Crane Lake P.O. as part of the regular winter routine. Miss Elizabeth, as we knew her, was a particularly pleasing keen-eyed young lady, proficient with a rifle, pack, or paddle . . . the certainty and alacrity with which her team of big, intelligent police dogs obeyed her every command was a relevation and an interesting performance to behold. The *American* magazine once featured Elizabeth as a comely and competent girl guide on one of their page portraits of unusual people . . . we probably derived as much pleasure in coming unexpectedly on this bit of national recognition as any regular in the Berger clan.

There must be some deep, fundamental, underlying reason why one meets so many interesting, friendly, and hospitable people in canoe land. They may be natives and rooted to the soil or they may be visitors just passing through, but surely there is some magnetic influence, some enchanting spell in the north country that draws such real folks together, that shapes their thoughts and very souls. It must have something to do with the breath of pine-scented air, the sparkle of sunlit waters, the cleanliness of rain-washed rocks, the beauty of a boundless woodland. No wonder men speak so softly and solemnly and reverently at times and say, "This is God's Country!"

NEVER THROW STONES
AT A MAYMAYGWASHI

Michael Furtman

No matter how many times we pass the pictographs on the painted rocks, we can't do so without stopping to admire them. Two of the most common questions we field are "where are the pictographs?" and "what do they mean?" I can answer confidently only the first.

The display on the high cliffs of Crooked Lake, just a little over a mile from our cabin and the falls, are some of the finest anywhere. At this place the rock leans forward and out over the river, leaving protected areas that were the canvas of the early artists. This cliff has been known for hundreds of years as the painted rocks, but strangely, not because of the pictographs. It has many vegetable and mineral stains as well as the plentiful lichens of the canoe country so that it is remarkably colored with sizable streaks of red, yellow, white, rust, and black. The actual paintings are small and unobtrusive, low along the water and mostly along the southern end of the cliff.

Much is still to be determined about the true nature of these haunting paintings. No doubt, as in the art of our own culture, they were painted for more than just one reason. There is evidence to suggest the more abstract pictographs might be the expressions of dreams as seen by the medicine men of the Midewewin society. Hunting magic, wherein the portrayal of your intended quarry helps to visualize the hoped-for outcome of a supply of needed meat, is common throughout the world in primitive cultures and it is quite likely that the Ojibway, or their ancestors, may have practiced similar rituals. Undoubtedly some pictographs are simply historical billboards, marking the passing of some important event.

But there is always an element of magic. The Great Lynx, or Misshepezhiew, can be frequently seen and is recognized by the horned head and long reptilian tail. Legends say that it is the Misshepezhiew that created storms on lakes by thrashing its monstrous tail and formed the shifting currents that capsized canoes in rivers. Many Indians offered gifts of tobacco to the Great Lynx to appease its spirit, especially before undertaking a dangerous, open water crossing or at the top of suspicious rapids. What better place then for an offering to the Misshepezhiew than at the confluence of confusing Crooked Lake and the perilous Basswood River?

Some passersby have informed us that though they paused to look for the pictographs, accurately describing the right cliffs of Crooked Lake, they could not see them. Perhaps they had expected something too large or too colorful, for the pictographs are small and dark red. Near the water, they are rarely any higher on the cliffs than a person could paint while standing in a canoe. Maybe those who are unable to see the pictographs are not just unobservant but are so civilized that these magic paintings simply are invisible to their eyes.

There is also a small figure here, off by itself, a tiny manlike drawing with arms outstretched. I used to call this little man the "first fisherman" since with the arms widespread he seems to hold the pose of so many anglers who brag about the one that did, or didn't, get away. I have since learned that this figure, which we have encountered in so many places in the northern lake country, is a Maymaygwashi. The Maymaygwashi is a figure of deep magic, both mysterious and sometimes devious. I once discovered just how devious he could be while canoeing the north part of the Quetico.

The day was a rare one, full of canoe country magic. We could feel this deep magic, feel its rhythms and mysteries as tangibly as though they were rain or wind as we paddled along. Caught up in the spell, I thought of those who had gone this way before us, those aboriginal peoples who had lived in this fair country. The ancients were in tune with those mysteries, understood the lake country's moods and had left sign of their understanding through the pictographs still visible on the cool rocks along dark shores.

We, who venture into this ancient land on occasion only, struggle to read the meaning in these paintings. We laugh at eccentric moose smoking pipes or the funny little men with arms outstretched. In the egocentric way of modern humans, we assume that we know more than our ancestors and that they were but children compared to us.

Though we understand so very little about the hidden meanings of these mysteries, we are no less affected by them as were those who formed those eloquent paintings. They speak of the elements, of wind and rain. They speak of food and war, of blood and wonder. They speak of man's need for a spirit world. These are things that none of us are separate from and though we might be sheltered by the cocoon of our civilized lives, all these things can, and do, affect us intimately and no less profoundly.

The day was warm and fair, blue skies a velvet backdrop for the few billowing clouds. A steady but manageable wind blew at our backs most of the day and we made wonderful time as we paddled down the lake's length. I kept a fishing lure dragging behind the canoe and we consistently caught fish, finally keeping one medium-size walleye for our dinner.

All down the lake's north shore we saw many pictographs. We stopped a few times to wonder at them, to take photographs. More than once we encountered the stick man Maymaygwashi and I tried to recall what I had learned of him.

I had read that the Maymaygwashi is a mischief maker, the Indian leprechaun. I had also read that the Indians believed these little hairy-faced people were fond of fish and often stole them from the Indians' nets, that they could change form and frequently played mischievous tricks. It is said that they could disappear by paddling their canoe into cracks in the cliffs.

Why had these little men come to be painted in just these spots? Was there some significance? Had there been an encounter with the Maymaygwashi nearby? We floated beneath them and wondered a while before moving on.

We selected a high, jack pine–studded island that evening for camp. Though the island was small, it afforded an excellent spot for our tent near its peak with spectacular vistas in three directions. After the busy routine of making camp, and before dinner, we settled down for a sip of brandy and the chance to write in our journals. At one point I glanced up from my writing for a view of the lake and saw a strange sight.

In the open water, about seventy yards from the camp, bobbed a fish on the surface. I thought this a bit out of the ordinary and watched it for a few seconds before returning to my journal. Then it dawned on me. Perhaps our walleye dinner had torn the stringer free.

Racing down to the water's edge, I searched the bushes for the stringer. Finding it still fast to the limb, I pulled it from the water. The walleye was gone! What was more confusing was that the cord stringer was still passed through the brass ring. The fish would have had to break its own jaw to have escaped. I thought that hardly likely.

I could see our dinner floating off shore. Sliding the canoe quickly into the water, I paddled furiously out to where the fish struggled, tail up. Moving in very quietly so that I would not scare the fish away, I eased my hand over the side of the canoe and as I drifted near the fish, made a move toward the now submerged walleye's tail.

As my hand had approached the tail I had seen that something was wrong with the fish's head. Indeed, something was very wrong. Hanging on to what was left of the jaw was a huge snapping turtle, one perhaps of forty pounds. He eyed me malevolently from two feet down.

Fully aware of what those jaws could do to my hand should he decide to strike, I nonetheless was reluctant to give up my fish dinner. Waiting until the snapper let the fish bob to the surface, I made a swift grab of the walleye's tail.

A tug of war ensued that I should have no doubt lost had the huge turtle not torn the head from the walleye. I quickly hoisted the remains of the fish into the canoe. In good shape except for the missing head, we would have our walleye dinner after all. I laughed at the turtle and paddled away.

With the weather warm and the fish now dead, it would have to be filleted and cooked before it spoiled. Dinner would come early. I knelt down on the sloping rock shore and began to fillet the fish on a canoe paddle. The sun was starting to drop and I could feel the warmth of the huge orb beat steadily on my back, a breeze cooling the perspiration on the back of my neck. Halfway through the task of cleaning the fish I had a strange feeling that the sun was not the only thing behind me.

Turning quickly, I was startled to see the big snapping turtle only a few feet from me, his back out of the water and his head only a few inches from shore. I jumped up and stood between the turtle and the much-desired fish. Do snapping turtles ever come out of the water? I was determined not to find out.

I found a large stone and hurled it near the fish thief, forcing his retreat to slightly deeper water. I hurriedly finished my filleting task (this time facing the water), rinsed the fillets, and headed to camp. I scraped the fish remains from the paddle to a rock on the water's edge to see if the turtle would come up and finish them off. If not, I'd return and bury them later.

There seemed something strange about this turtle's conduct. The thought of it pestered me.

The fish dinner allowed me to forget my concerns. A fine, calm evening settled in, just made for a leisurely paddle. Filled with the sounds of dusk and the swishings and swoopings of nighthawks and bats, we hurried the after-dinner chores so that we might have time to explore before dark. Leaving the cleaned dinner dishes overturned on a rock to dry, we launched the canoe and drifted into the last light.

In the twilight, Mary Jo asked to be put ashore for a few minutes. We found a likely spot along a dark, spruce shore and I paddled the canoe parallel to the sloping rock. Mary Jo and Gypsy hopped out and disappeared into the bushes. For a moment I sat calmly enjoying the evening.

The next thing I knew I was submerged in five feet of water, the canoe floating over my head. I quickly found bottom and stood up, thrusting my head into the air. I looked around, startled, and saw the canoe floating right side up, high and dry, a few feet from where I had let my wife out. Heavy with water-soaked clothing, I dragged myself turtlelike onto the rock shelf. By this time Mary Jo had returned, and relieved that I was OK, began to chuckle.

"Did you see the turtle down there?" she asked, laughing.

The turtle! Of course, the turtle! Somehow he was behind this, somehow he had managed to get even. But could a turtle be so scheming? It was then I recalled the pictographs we had seen on this lake today. We had seen so many little men with arms outstretched! I quickly recalled their mischievous ways, their extreme fondness for fish. Whether from wet or wonder, a chill gripped me.

"Well," repeated my wife as she steadied the canoe for her soggy husband to step into, "did you see the turtle?"

"That was not a turtle," I said.

"Oh? What was it?"

"A Maymaygwashi."

"A what?"

"Maymaygwashi. You know. The Indian leprechaun."

My wife chortled in comprehension. "Well, of course, dear. You should know better than to throw stones at a Maymaygwashi."

"Hmmphh." Indeed.

I have pondered this event in the days that have gone by and have become thoroughly convinced that my surmise was correct. Surely I had met a Maymaygwashi. To me there could be no doubt. And at last I began to understand that those who lived in the canoe country in years past knew well what they were expressing when they painted those pictographs. If I had had some red paint on that canoe trip, I too may have left a Maymaygwashi warning on some rock cliff near where we had matched wits.

Just what had happened near the cliffs of Crooked Lake to be responsible for the appearance of the little stick man on those rocks is unknown. But I feel as though it must have been something that was otherwise unexplainable, and perhaps even humorous. That is the nature of the Maymaygwashi.

If you ever paddle near there, I'd be damn careful.

A DOG TEAM TRIP

Justine Kerfoot

In 1931 I invited Gene Bayle to join me on a week-long dog team trip to visit Art Smith at his trapping and hunting cabin on Mountain Lake. We packed the toboggan carefully with sleeping bags, food for ourselves and the dogs, and the necessary cooking pails. The dog team toboggans were long, narrow, and limber, which enabled them to fit a snowshoe trail and snake over uneven ground. We started with a speedy dash, but the dogs soon settled into a steady trot. As we approached Charlie Olson's cabin down Gunflint a little way, we saw him on the lake getting a pail of water. Although we had barely started, he insisted we stop for a cup of coffee.

Charlie was active in spite of his seventy-five years. He skied up to our lodge at least once a week, cut his own wood, did his own cooking, washing, and chores. He lived alone with his dog. He was born in Norway, in a region that looks very much like the Gunflint area, and he retained his Norwegian accent. For years he was a logger in the northwoods camps. Many fellows like Charlie are rugged, gruff, and painfully outspoken, but under their hard exteriors they are kind and generous. They remind me of an old Indian legend which says that no man can live among the pines who is not at peace with himself.

We stepped into Charlie's spotless cabin. The floor looked freshly scrubbed and the bed was neatly made. Charlie motioned us to sit at his table where he offered butter, crackers, maple syrup, and strong black coffee.

"You have a lot of birds feeding on your bird shelf, Charlie."

"Oh, yah! Two pair nuthatches, six pair chickadees, two pair small voodpeckers, und four pair large uns. De viskey yaks steal so much. Ay chase dem away. De nuthatches coom to me every year. De voodpeckers eat such a lot, soon I'll have to buy more suet."

"What shape are the deer in, Charlie?"

"Oh de deer are awful poor. Ay've been cuttin cedar for dem every day. De game vardens promised me hay but ay ain't got any yet. Last night five deer coom down in de yard. Split Ear is back again dis year. Ay seen him yust last night ven ay was cuttin wood."

"What are you going to make with the diamond willow and cedar you have drying?" we asked.

"At tank ay make a couple of chairs wit de cedar and a sideboard wit de diamond villow. Oh it's a lot of vor-rk. Ay had a hard time finding de diamond villow last fall. Looked all over to get some goot pieces. Vere ya goin?"

We answered, "Down to Mountain Lake, going to be gone almost a week."

"Yah! Vell yore dogs look in good shape—dat team dat vas here yesterday vas so poor ya cud see tru em like paper. Vell ya haf a long vay, ya better get goin."

We pulled on our parkas, hopped onto the toboggan and started down the lake—the dogs traveling fast. The sun transformed everything into a dazzling whiteness. The only shadows were those of the dogs, bouncing along easily on the glistening blue-white snow, followed by two shapeless and practically motionless forms—Gene and me on the toboggan.

Far down the lake we saw a dark object on the ice that appeared to be a stump. As we approached, it took the form of a fox sitting on its haunches. Because we were up wind, the fox wasn't startled. We came within fifty feet of him, then the fox started to lope lazily in a half circle. The dogs wheeled fast at full speed. Suddenly the fox got a whiff of us and became a receding streak, zigzagging across the ice. In a moment he was all of two miles up the lake, then he disappeared behind a point and was gone.

In a short time we reached the inlet where Little Gunflint flows into Gunflint Lake. The current here can make the ice unsafe, so we followed the old railroad spur on land for a short way. About halfway over this portage we stopped to cook up. We unpacked our tea pail, in which we had a little tea, sugar, cups, and a couple of sandwiches. While Gene went down to the mouth of Little Gunflint for water, I kicked around for some dry poplar, often called squaw wood, which makes a quick hot fire.

I cut a green tag alder, trimmed the branches, stuck the heavy end in the snow and angled it to hold the tea pail. I gathered a handful of birch bark and had the fire going when Gene returned. We hung the pail on the tag alder over the fire and went back into the woods to gather balsam boughs to sit on.

While gazing pensively into the fire and waiting for the water to boil, we remembered a poem by Bliss Carman in *Songs from Vagabondia*:

> Here we are free / To be good or bad,
> Sane or mad, / Merry or grim
> As the mood may be, —/ Free as the whim
> Of a spook on a spree—

As we ate our lunch, the dogs eyed us with drooling anticipation, watching each bite disappear. We kicked snow on the fire and resumed our journey.

The dogs started out with renewed vigor after their brief rest. As we jogged past the rapids, the trees squeaked with their coatings of ice formed from the open water's steam in winter air. Delicate fingers of ice, more beautiful than etched crystal, extended from the shore above the rushing water. The snow on the rocks appeared to be covered with a pure white fuzz. On closer examination we saw tiny clusters of ice perfectly shaped and as exact in design as snowflakes.

As we pulled out onto the lake ice, again we saw a dark object on the ice ahead of us. It looked like an old blanket that had been tossed aside. As we came closer we recognized the hide of a deer—skinned as neatly as if done with a knife—and a few bones, surrounded by wolf tracks.

We drove out onto North Lake, a six-mile white expanse bordered by high dark hills. This lake, like most in the area, lies east and west with the higher, more abrupt cliffs on the south side and gentler slopes on the north. We skirted the lake on the northern shore, taking turns trotting after the toboggan to keep warm.

Joe Blackjack's cabin was located halfway down the lake. He spoke very little English but he worked for me for several years with the willingness and strength of an ox. I have watched him shoulder a log eight feet long and walk off with ease. His wife once gave me a woven rug made from strips of the inner bark of cedar trees, colored with dyes made from boiling woody plants.

Joe lived with his wife, children, and innumerable dogs. This Indian family, different from our closer neighbors, lived in the crudest manner without much cleanliness. When we approached their cabin, a chorus of howls greeted us from a motley pack of nondescript dogs. The door opened and Joe came out to greet us. Behind him, framed in the doorway, four curious children were peering out. Joe, dark, swarthy, stockily built, and very bowlegged, was a full-blooded Chippewa. His open shirt revealed a bare chest. His unkempt hair and the general appearance of his face and hands answered any doubts about his personal hygiene. His children substantiated the same thought.

Judging by the appearance of the walls and roof, their cabin barely hung together with the family dependent on their own huddled numbers for warmth in cold weather. There were no windows to give light, and a dirt floor was within. To Joe, his wife and six children, it was home.

A couple of Joe's well-placed kicks cleared the dogs to one side as he came down to meet us with a big grin of welcome.

"Bi-jou."

"Bi-jou, Joe. How's everything?"

A happy grin and a shrug of his shoulders were followed by, "Trapping no good." That told us that he and his family were well but that he, as usual, was broke and about out of chuck.

"Got snuff?"

"Sure, Joe," I said, and handed him two boxes that I had brought for him. He opened one box at once and helped himself to a generous amount. Then he turned and tossed the other box toward the doorway. One scramble and the children had opened the box and partaken of its contents.

"Where you go?"

"Mountain Lake, Joe."

A cheerful grunt, then a wave of his hand was our parting signal. We headed for Sac Bay where there was a hard trail following an old logging road to the Rat Lake portage. The spruce and balsam, so heavy with snow, almost arched the trail in a white-and-green bower. Here we were protected from the wind and had a feeling of snug warmth. Where the white-draped branches of a tall pine ended, a snow-covered balsam took over producing a featherbed slope down to the ground.

After the portage we came to Rat Lake—not much more than a beaver pond separated from Mud Bay by a narrow spit of land. On this bit of land was an old trapping shack that Art Smith used occasionally, so we decided to stop and look inside. Art had told me once that he usually left a little chuck under an old trap door in the floor. There was a tiny airtight stove, a pole bed covered with balsam boughs, a table, a couple of fox stretchers, several mink stretchers and in the far corner was a homemade mouse trap. A bit of kindling and chips were piled beside the stove. I spied the trapdoor and pulled it open. There in a shallow pit were three cans of beans.

Next we drove down Mud Bay and on to Rose Lake. High cliffs towered above us on our right. Bits of moss topped with snow clung in tiny crevasses to the otherwise bare cliff face.

A little beyond was "stairway portage" to Duncan Lake—a series of long steps embedded in the earth. A small stream tumbled down the cliff in a series of little falls on one side of the portage. On the opposite side more cliffs guarded the entrance to Arrow Lake. An old Indian legend says the Indians competed here in shooting arrows to the top. We continued to an abandoned railroad bed and then turned onto a steady incline for about a mile to Daniels Lake, where there was another trapper's cabin.

We arrived just as the sun was dipping behind the hills, and the evening chill was descending. We unhitched the dogs, tied them to trees, and gathered some balsam boughs for their beds. Then we unroped the canvas covering our gear and moved in. In the sky to the west color streamers were flying. The setting of the sun was followed by long black banners unfurled by the wind, then the shadows blended and night came.

This cabin had two double bunks of fresh balsam, a small airtight in one corner and a table by the window with two stumps for stools. In a short time we had a fire going. We threw our eiderdowns on our bunks and started supper. I mixed a little bannock dough consisting of flour, baking powder, salt, and canned milk. Bannock resembles baking-powder biscuits and is a good substitute for bread. The dough is flattened out and fried until it rises and turns brown. Then it is flipped and topped with brown sugar.

After supper, with the dogs fed and dishes washed, we sat at the table. By flickering candlelight we looked over an old magazine we flushed from a corner of the cabin. I tossed an extra chunk of wood into the airtight and it began to sing. The wood was alive again, once more standing tall in the forest, its sap running up to its branches and the soft south winds creeping through the leaves. Like a record playing for the last time, the wood told of life among its fellows. Then it crumbled to ashes— source of nourishment for another tree.

We blew out the candle and climbed into our eiderdowns. The warmth of the bags, the scent of the balsam boughs, the long squeak of two trees as they rubbed together outside the cabin, and the tiny rustle and gnawing of wood borers working under the bark on the cabin logs all added to our contentment. A light from a crack in the lid of the airtight flickered on the ceiling, and soon even this disappeared.

The next morning was considerably warmer with a bright sun. By this time Gene and I had evolved a natural sharing of work. Gene set the table and got fresh water while I cooked breakfast. She did the dishes while I replenished the wood supply and repacked the toboggan. We hitched the dogs, took a last look, and were ready for our trip to Mountain Lake. We followed Art's trapping trail, crossing a little stream that was the boundary between the United States and Canada. Tracks indicated a mink had followed this route in search of food.

This old trail from Daniels Lake to Watap Lake has been traveled by hundreds of couriers and Indians packing out fur in the spring. It has felt the soft pads of innumerable dog teams. The trail follows a draw, well protected by hills, where the sun only penetrates in small patches. We entered a tag alder swamp just before reaching Watap Lake. As we came onto the lake we stopped and made birchbark goggles with just a slit in front, to eliminate glare from the snow and prevent snow blindness.

The morning was balmy with a sniff of spring in the air. As the toboggan slipped along, we threw back our parka hoods and pulled the liners out of our mitts. As the sun rose higher, we shed our parkas and pulled out one set of shirttails. (We wore two wool shirts one over the other.)

Halfway down the lake we noticed a live beaver house along the shore. Beaver have their front door underwater but they live in their house above the water line. In front of the house the ice is always thin from their underwater travels. Anyone trespassing too closely is assured an icy bath. In winter beaver trappers make their sets under the ice. A good trapper will take only a few beaver from each house, preventing the depletion of his own trapping grounds. In spring one can hear the little beavers a'talking in the house.

It was a short run to the portage—up the hill, down the other side, and out onto Mountain Lake. The lake stretched ahead for six miles, guarded on both sides by high hills. The heavily wooded hills slipped by, dark in their wild secrets—a woods dense and at times uncompromising. We could hear the soft rustle of the snow as it whooshed off the fir trees. The branches, released from their weight, bobbed up and down as if trying to restore their circulation. We rounded the last bend and saw fresh smoke puffing from the cabin chimney.

As we approached the cabin Art opened the door, leaned against the door jamb, and asked,

"How's the going?"

"Good," we answered, "we came right along."

"I seen Harry when he came through so I knowed you would be along. Tie your dogs out to those trees and I'll yank the toboggan in the cabin and unload."

The cabin was littered with wood chips. Art was in the process of making a runner for a new sled and drying slabs of cedar to be shaped into paddles. In one corner were some burls from birch that he would make into bowls and cups and saucers. Some fellows have the knack of making about anything out of wood, and Art was one of them. Even the wooden door hinges he had made were functional. He remarked, "I've been working on this sled; the parts are about finished. All I need are a few bolts to hold the parts together. This sled will be light and good for traveling in the spring. I fixed it for you, you can take the parts back, get some bolts, and sometime when I'm up that way I'll help you put it together."

"How's trapping, Art?" we asked.

"There's not much. No goddam game left in the country. It ain't worthwhile for all the running around you have to do. The country is about ruined, too many people. Fishing ain't much good anymore, no marten left, hardly any fisher or otter, only a few mink and weasels and every time you look for a beaver the goddam game wardens are snooping around. There ain't any game left in the country at all. When the logging camp was here, I was hired to shoot moose for the camp, and that was easy. You couldn't do that now; even the moose have moved out. There ain't no caribou, and there used to be some of those up here. There ain't enough game in the goddam country to make a good living from anymore. Think I'll go further north where there is still some fur."

It was late so we busied ourselves getting dinner. Art stepped to the door, reached up on the roof where he had cached a moose heart for dinner. He cooked the meal while we went to the lake to the water hole. This one was covered with a bottomless box, well packed on the sides with snow, with a lid to deter a fast freeze. By the time we were back and carried in a few armfuls of wood, Art had the meal on the table. After cooking for the dogs and washing the dishes, we watched the long afternoon shadows retreat step by step and vanish on the far side of the lake.

Art said, "I've got a trap down here about a mile; ain't seen it for a couple of days. Want to put on your snowshoes and come along?"

"We're with you," we answered. We hiked down the lake to a live beaver house. Art chopped a hole in the ice, pulled out the trap with a beaver in it, and tossed both of them in his pack. He was taking no more beaver from this house for this year.

We returned to the cabin and watched while Art skinned the beaver. Deftly he slit the belly from the tail to the tip of the nose. He made four circular cuts, one at each wrist so later the feet could be pulled through the apertures, and then skinned the animal in just a few minutes. The hide was clean, golden in color, and showed no evidence of grease, which if left could cause a "burn" when the hide dried.

He picked up a hoop he had made by bending, twisting, and then tying a couple of young ash sticks together. He threaded a large, curved hide needle with light meter line and sewed the hide into the hoop. He stretched and sewed at the same time, creating a "blanket," which was slightly oblong but had well-rounded sides. After he completed the work to his satisfaction, he put the hide in a cool corner of the cabin. In three or four days the hide would be dry. Then the fur could be brushed and the hide rolled into a loose bundle for packing out later.

The next morning continued warm and we busied ourselves, under Art's guidance, making wire snares for rabbits. Gene took four and I took four, and we headed for the nearby patch of green timber where there were a number of trails. We picked places where the rabbits traveled under windfalls. We shoved down sticks on each side of their trail, leaving just enough room for them to travel through. Then we hung our snare in the center of the runway a couple of inches off the ground. It took us some time to find the best spots and to arrange everything just right. We looked back, as we returned to the cabin, in anticipation of a rabbit stew.

Gene and Art began whittling small paddles from pieces of dry cedar, while I busied myself making first a holster for my

.22 Woodsman Colt from an old piece of leather boot and then a knife sheath from a beaver tail. We were an industrious group.

Suddenly Gene stopped whittling and said, "You know a vacation in the bush can put a person's feet back on solid earth, renew perspective, and give one a dioramic outlook on life when the surface has become flat without light or shadow."

"H-u-m-m-ph," Art grunted. "The city is no place to live. Just a jammed-up bunch of people, one living on top of the other. They talk so much about lakes that they have fixed up. Why hell, they ain't as big as one of these damned old beaver ponds. They ain't got no fish, except for a few about as big as yore finger—then they go out and set all day to catch those. No game except what they have tied up in their parks or stuffed in their museums. Oh I've been down there—everyone running around in such a hurry, cars banging up and down the street, people milling in and out of the stores. Why, I bet half of them don't know where they're going or why they're in such a big hurry. Then look at all the smoke and dirt hanging over one of those places—you can hardly breathe, let alone see the sun—people setting in offices all day with lights on—hell, that ain't livin'—it ain't even hardly existin'. Nope, this is the kind of country to live in, only there ought to be more game so a man could make a good living without having to run all over the country for one lousy mink or weasel."

Art's paddle was swiftly taking shape when he said, "There ain't much good cedar around. You look until you find a tree where the bark markings are straight up and down. Most of them go around the trunk and they aren't straight-grained. You have to cut the length you want and then bring it in to dry. You dassn't use the center because the core ain't no good but you use a slab, free of knots, alongside the core. Sometimes you can find a dry cedar that the Indians have stripped aways to make bark baskets. You can tell them by the checks in the tree. Lots of times in trees you find like this wood has lost its life. It is usually good enough for making ribs for a canoe."

In midafternoon we decided to check our snares to see if we had a rabbit for supper. As we stepped out of the cabin the sun was sliding down over the hill. Gene and I were both a bit disappointed to find our snares untouched. We cooked up and after eating climbed onto our bunks, comfortable and content. We planned to start home the next day. Art would go with us as far as the Daniels Lake shack where he would turn off and go to Clearwater to get more food.

Suddenly out of the darkness came a shriek like the shrill cry of a child. We dashed out to our snares. Sure enough, there was a rabbit. In our chagrin for causing such pain we quickly took him out of the snare, but he just lay there on the snow in the clear moonlight mortally hurt, gazing at us with his big soft eyes. We felt sick in spirit and didn't much want to have the fellow for a stew. Art looked at us with an understanding smile. While we picked up the rest of our snares, Art cleaned the rabbit and set it on a rafter for a meal upon his return.

The following morning there was a slight snow flurry. We spent a couple of hours around camp, packing up and gathering a good stockpile of wood for the cabin. The snow was coming down a little faster by the time we were ready to start. Art snowshoed ahead, and as we drove out of the bay and around the point, the fury of the storm struck us head-on. Art stopped and said, "It looks as if it might be tough."

We proceeded down the lake, and the storm became a seething, swirling mass of snow. It was like a thick fog obliterating all the shorelines and leaving us with no sense of direction. John, the lead dog, stopped and looked back. We got off the toboggan and put on our snowshoes. I went ahead to break trail while Gene followed the dogs. It was almost impossible for me to follow Art's windswept track although he was only a few yards ahead.

It was, as Art had predicted, tough going. The hoar frost from our breath froze on our chins and the fur of our parka hoods. Snow gathered on our eyebrows and eyelashes. The dogs' faces were encased in white. Just dangling tongues and sets of eyes indicated they were dogs. I turned to see if Gene were following. With my back to the wind I thought how easy it would be to travel with the storm, like coasting downhill. As we pushed on, our pace slowed. The toboggan was dragging hard and the dogs had begun to tire. We moved like robots leaning into the blizzard. At times the snow swirled deep into our noses and throats almost restricting our air supply. For a second the storm lifted, and I saw Art standing far ahead of us and over on the shoreline. We headed in that direction, but all was blotted out again. In fifteen minutes or a half hour or maybe more, we pulled into a tiny sheltered bay. Here there was no wind, just a fine sifting snow coming down through the trees.

Art looked us over carefully and then said casually, "We're about halfway. I believe we'd better pull into the swamp here and camp for the night. It's almost three o'clock and we can't make it in against this storm before dark."

The cedar swamp was well protected, and soon we had a fire going. It was a relief to get out of that driving snow.

"I'll fix a lean-to over here and the two of you get a big bunch of balsam boughs," Art said as he shoveled snow with his snowshoe. He built a shelter and carefully placed the boughs we had gathered on top like a thatched roof. We piled a thick layer of boughs on the ground inside the shelter and had our house completed. Art dragged in enough wood to last through the night while we bedded the dogs, unpacked, and melted snow for coffee and cooking. After supper we sat inside our shelter on our two eiderdowns, enjoying the warmth of the fire and protection from the snow. At Art's suggestion we clipped the two eiderdowns together, removed our parkas, outer shirts, and moccasins and all slipped into the enlarged sleeping bag. From far off we heard one long mournful howl of a timber wolf. Egi, the malamute, lifted his snoot to the sky and answered. We slept the deep sleep of the weary.

Sometime in the night or early morning I became acutely aware that the storm had abated, the fire was out, and it had turned very cold. From then on we slept restlessly, not cold and yet not warm—just on the margin of comfortably uncomfortable. By daylight the trees were snapping with the cold, the top edge of the eiderdowns was covered with hoar frost from our breath, and the dogs were curled into tiny white humps. We drew lots as to who would get out to start the fire and cook breakfast. I pulled the unlucky draw, and as I crawled out into the numbing cold, I wasn't particularly pleased with the smug looks of satisfaction on my companions' faces.

Breakfast was corn fritters and coffee. The snow squeaked underfoot as it does in intense cold. The fritters almost cooled while cooking, which led me to believe it was colder than forty below zero. By the time they landed on our plates, all warmth had completely departed. The coffee, however, did stay hot until it could be downed. We did not tarry long after this meal, as we felt we had to keep moving to keep warm.

In a surprisingly short time we had packed, hitched the dogs, and were ready to leave our comfortable quarters. We had Nookie on lead because he was trained to follow a breaking trail and never stepped on the tail of a snowshoe. John was moved back into the team; he dropped his tail, flopped his ears, and sulked. He still pulled, but only like one who was highly bored with a most unpleasant task.

Soon we were at the parting of our ways. Art went on to Clearwater while we turned back to Rose Lake and our return home. We took turns, riding and trotting, and about noon we stopped at the North-South Lake portage to cook lunch. Just as the tea water was boiling, George Plummer came around the point. He knew a shortcut to Gunflint and offered to show us the way. We traveled up an arm of North Lake, into a cedar swamp through which a very crooked trail had been cut, continued across a beaver pond, through more cedar swamp and out onto Gunflint. Mile after mile Georgie trotted beside us with no hesitation—a tribute to a good physique. We were opposite Charlie Olson's cabin as the curtain of darkness descended. The night was clear, crisp, and blue. Some of the brilliance seemed to come from below, crystals of snow winking back the starlight. The lake groaned and moaned as its bars of ice were strengthened. A person alone on such a cold night is glad for the friendly squeak of snowshoes on cold snow. But the two of us riding behind the dogs toward a familiar glow of light reveled in the feeling of belonging to a world apart.

An epoch of time passes—a historical interim—for trappers of this breed are no longer. These men of the woods would guide canoe parties in the summer but seldom a fishing party—only if they were fly fishermen. This handful of local men—Art Smith of North Lake, Gunflint, Clearwater, or wherever he hung his hat; Jimmy Dunn of Sea Gull; Jack Dewar from Loon Lake; John Clark at Gunflint; Benny Ambrose of Otter Track Lake—gave to any party they took on a trip an unforgettable experience and a feeling of comfort in the woods. People from all over were privileged to travel with these experienced and congenial men. They were replaced by college boys who became known as fishing guides.

CHRISTMAS AT BIRCH LAKE

Carl Gawboy

It was cold. There were great columns of icicles from the eaves to the snowbanks piled high against walls. There were brothers and sisters there, and heaps of cinnamon rolls, puddles on the floor, light, and noise.

It wasn't always that way, of course. Between 1946 and 1954 there were many Christmases. There were the spare Christmases of the hard years, but others when Pa had work. Sometimes the siblings all came home from college but sometimes they only called on Christmas Eve. There were mild winters, too, and one with rain. But in my memory, the Christmases blend to a Christmas that always was.

On the day before vacation, the sidewalk in front of Washington School thronged with children—galoshes flapping unbuckled (a forbidden fashion), wool coats, aviator caps, and clanking lunch pails, piles of papers from cleaned-out desks, bags of treats from classroom parties. The White Iron–Birch Lake bus pulled up to its place between the one for Spaulding Location and the Tomahawk Timber Company bus bound for Forest Center. As the bus door opened, children large and small elbowed for a place in line. Shorty Lenich sat at the wheel, beside a great cardboard box. As each child stepped in, Shorty handed him a tinseled box covered with images of Santa Claus and Christmas toys. Inside was candy of all stripes and colors. The bus filled and children stood in the aisle from front to back. It smelled of wet wool, leather, rubber, and candied breath. The windows went opaque with frost, and those lucky enough to have a seat made imitation footprints with their clenched hands. Woe to anyone who dropped a book or paper on the floor. Melting snow from boots and snowpants made the floor a cold winter stream complete with ice chunks and mud.

The bus pulled away from the curb and headed south on Highway One. Each stop relieved pressure. The Rodich, Weijo, Shroeder, and Kisrow families all departed with their bundles of treasure. When Chubs Johnson vacated his seat, we gathered around, marveling at the great crater his bulk imprinted in the leatherette. (I remember that pit remained as his memorial long after he graduated.) By the time the Esterbergs, Dojans, and Starkmans departed, the aisle was clear and everyone remaining enjoyed the luxury of a seat.

The bus turned onto the Babbitt cutoff. After the Korppi place was the long stretch of highway past the Buski farm, people too mean to have children. The winter night had come by the time we rounded the curve and climbed the hill to the Birch Lake Road. A circle of mailboxes mounted on a wagon wheel was our stop. The Bischoffs and Jon Wilmunen got off with us. The Tuomala kids were already moving into our seats as we ran down the trail shortcut that Jim kept tramped along the old tote road.

We thundered into the house with our booty. It was chore time, and Jim and Ma went to milk the cows. My sister Pat and I hauled water for the cows and for Nellie, the old horse. We forked hay and bedded them down for the night. We carried wood by the armfuls into the house. Pa came in from the woods and it was time for supper.

From the oven of the great black kitchen range, Ma pulled out the large four-loaf bread pan topped with mounds of golden biscuit. A dipper broke the crust and ladled out chunks of venison, carrot, potato, and rutabaga swimming in a dark brown broth. The older kids wedged around the kitchen table while Pat and I shared the woodbox. We cast a covetous eye at the table. When Jim went off to college we would sit at his place and eat like grownups.

In the days before Christmas, Ma drove the jeep to town, meeting trains and buses as the older siblings came home for the holidays. Irene and Bob taught us the Harvard fight song and "Hail Our Dear Cornell," even though they attended Milwaukee Downer College and the University of Minnesota. They also sang off-color lyrics and told jokes daring us younger children to understand. They swaggered, these college siblings. They talked like college people talked in Donald O'Connor movies. Jim and Pa brought home a tree wedged on the top of a load of logs old Nellie pulled in with the bobsled. The tree was hauled in with pomp and ceremony. One year we bought our first lights. We hung them on the tree first. Extension cords of various gauges connected into our light socket in the middle of the ceiling. The magic moment arrived. The lights came on. I nearly fell over backward with the loveliness of it all.

"Wait, wait," Ma said.

"Wait, wait," Jim said.

Each light had a tube of glass filled with a colored liquid. A bubble rose in one.

"Look, look," Ma said.

"Look, look," Jim said.

Soon another bubble, then another. The whole tree was filled with light and glass tubes bubbling away. I embraced the wonders of the modern age.

Pat and I believed in Santa Claus, of course. We knew he came over the back field from the Tuomalas', and we knew that at our house he came in through the door. Ma and Pa had known him for years, and he would visit with them in the kitchen and drink coffee. We knew we must never interrupt their visit, not even for a drink of water.

Pat told me that she saw Santa Claus once. She was sleeping in the top bunk next to Irene. She heard the jingle bells and listened to the voices. She peeked through the ventilator hole between the two rooms and saw the tree in all its glory. There was Santa, setting out presents. He threw his head back and laughed heartily. Pat had never told anyone about this before and she told me I must keep this terrible secret. I have kept it—until now.

Most of the time our house was filled with clutter. There were books and magazines piled everywhere. On winter days, sheets hung dripping from clotheslines strung across the living room. Papers, games, school projects, guns, traps, and an occasional muskrat carcass in the process of being skinned filled the small room. But on the days before Christmas our house got a thorough cleaning. All the girls were put to the task. With kerosene-soaked rags they dusted the stove, the tables, and all the woodwork. The men were kept out of the house while the floors dried from their scrubbing. The rag rugs, beaten and laundered and fresh from the out-of-doors, were laid down.

Ma and girls swung into action in the kitchen. Pans of fruitcake, great golden cinnamon rolls with candied bottoms, doughnuts deep-fried in oil, and wild rice candy, made by popping wild rice like popcorn.

In the evenings we read or drew, listening to the Christmas episodes of our favorite radio shows. Gene Autrey and his Melody Ranch gang always did "Santa Claus Is Coming to Town," and I wondered why no one applauded in the Old West. After a song the cowboys just said, "That was real nice, Gene." On "Amos and Andy," Amos would tell the Christmas story to his little niece as she fell asleep. I was very impressed when the announcer told us that the show was beamed to our Armed Forces around the world. The world seemed secure, protected: and those protectors could gather around their radios and feel good, too. Gabriel Heater and Edward R. Morrow brought us the news. Lionel Barrymore played Scrooge in "A Christmas Carol." I wasn't happy with Marley's ghost. I didn't think a Christmas story should scare little kids.

Christmas Eve was the most memorable of nights. We took our saunas, two at a time, and made the dash through the porch to the house. Jim and Bob slept upstairs under piles of quilts, where the arctic wind blew through the walls. The girls were in the bunks, two to a bed. Ma and Pa stayed up late reading or talking. I lay awake on a mattress on the floor, thinking about my world. We all were clean, the house was clean, there was a fire in the stove throwing its flickering light on the wall. In the barn, I knew, the horse stomped, the cows munched, and the chickens chirped sleepily on their roosts. Outside in the night sky the magical star of Christmas shone over the spruce and birch woods, over the Wilmunens to the south and the Tuomalas to the west. And somewhere in the arctic gloom was the Jolly Old Elf, winging his way to Birch Lake Township.

We children got up first, and gathered around the tree to gaze at the presents. Ma lit the fire, put the coffee on, and started the oatmeal cooking. The presents were unwrapped. I got a ship that exploded with a wooden torpedo, or a toy truck with a picture of the driver in full face on the windshield and in profile in the side window. Jim and Irene told me gravely that when they were my age all there would be for toys were two blocks of wood nailed together. How I wished I could have been there, during the Depression.

The best gift I got, and one I always received, was a full box of sharp new crayons. Sometimes it was a forty-color set with colors like Flesh and Veridian. There was a ream of clean white paper to go with them.

With the opening of the presents, the clutter miraculously reappeared.

Christmas dinner consisted of an entire venison haunch roasting in the great oven in the wood range, mounds of mashed potatoes and rutabagas, hand-parched wild rice, home-canned pickles and beets, cranberry sauce, and rolls and bread fresh from the oven.

After dinner, Nellie was harnessed to the bobsled, and off we went to bring back hay from the hayshed on the other side of the farm. We made a great load, the older kids pitching the hay with Pat and me tramping. Nellie hauled it home and we forked it from the load into the barn. (Hay scattered on snow always reminded me of milk curdled in rhubarb sauce.) Afterward, when we came back into the house, the older kids played Chinese Checkers and I colored. That is, I colored as well as I could with my numb and thawing fingers.

Then Christmas was over, and it was time to bring the older kids back to town, to catch their train or bus to the world outside. They seemed excited to go. Those older siblings had an arrogance, a sense of belonging to new culture, one that was alien to me. I felt that perhaps they were chosen for greatness, maybe the cover of *Time* magazine. This house on a rock and swamp farm was a backward place, a Dogpatch, compared to what was in store for them. They told me so. What didn't occur to me then was, if they felt that way, why did they come home? I didn't understand it until a decade went by and my world changed. I was at college in a bleak boarding house. My roommates from small towns and farms did the Donald O'Connor role in their school sweaters. It was then that the vision came to me: a vision of black spruce swamps, snowy fields, venison stew, a kitchen filled with smells of canning and baking, a father smelling of tobacco and pine pitch, and a short, round mother with flour dusting her apron and her arms.

A DAY'S CROSS-COUNTRY SKIING IN VOYAGEURS PARK

Robert Treuer

We each had our own motives and agendas. People usually do.

At the sporting-goods store in International Falls where I had rented cross-country skis and boots the night before, the owner chatted with me after sharpening a youngster's hockey skates and advising another on different ski waxes, some for cold weather, others for warmer days.

"I used to work for Boise Cascade," he said. "They wanted me to transfer to the main office in Idaho, and I couldn't bear to leave this country."

"The price of success?"

"I suppose. But it's hard to make a living here. I finally qualified and got a teacher's certificate, and am teaching school. My wife and I run this store, which I just love. You can say we stayed here because of the park. You going to ski in the park?"

"Tomorrow we'll drive to Ash River, ski up to Hoist Bay and around some of the Namakan Islands."

"It should be a good day for it."

It was the end of winter. There were still over two feet of snow on the thick lake ice. Melting and the breakup of the ice was still wishful thinking, but the fish houses were being dragged from the ice on the off chance that it would warm up any day.

"I've canoed through the border lakes and in the Quetico," I told him, "but this is my first winter visit here." I expounded on my private theory that short, bowlegged people such as I have an advantage over tall, long-legged folks on snowshoes. I had tramped over a proposed loop trail on the Kabetogama Peninsula of Voyageurs National Park that day with naturalist Frank Ackerman, who wanted to show me the difference between logged areas in the roadless, rocky terrain and some spots where bigger timber remained. Padding over the deep snow was a lovely way to get around, quiet and easy, like scuffing over the surface of life without disturbing what was underfoot. Until we came to deadfalls and thick tangles of brush. I had been confident about the expedition, having spent some time on snowshoes as a northerner, but a day-long cross-country-skiing expedition would be a different experience; I'd never tried the new equipment.

"You'll enjoy it," he said. "There are places out there you can't get to except on skis or snowshoes. Probably won't see another person all day, once you get into the park."

I looked dubiously at the skinny cross-country skis, so narrow compared to my old pairs of wood downhill skis. Now, how would I manage on these? I'd willingly go anywhere, match skill with anyone on snowshoes or in a canoe, but I had little confidence about the morrow.

"These things are quite in demand now," the hardware-store man assured me. "We've got people coming up here all winter long to go cross-country skiing. It's like they said, the park is drawing more people all the time. I'm switching to those skis over there next year." They were even narrower.

As I left the friendly store I wondered about the attraction of the north country. Here was a competent engineer who preferred to change occupations, taking a teaching job at much less pay rather than move away and transfer to a better position, just to stay near the northwoods. And it was not merely a matter of tolerating the long winter in order to enjoy the brief, brilliant temperate times. I shared this madness and could not explain it either, though my home was a hundred miles away in the rolling hill country of the Minnesota north, not on the stark, dramatic granite of the border lakes.

The morning dawned with a faint promise of sunshine amid uncertain temperatures. Maybe fifteen above? Twenty? It was a bit humid, making the cold penetrate clothes. Five of us gathered at the park's administrative offices just outside International Falls, longing for a bit more sleep and warmth but not admitting it. The park building was a modest, one-story wood structure, temporary housing until the permanent building could be erected a few miles away on Black Bay of Rainy Lake. Frank Ackerman tried bustling around and being efficient, gathering equipment, maps, and canteens, with indifferent success. Mary Lou Pearson, the International Falls schoolteacher who doubles as park historian, kept looking longingly at the warm building when she thought no one was watching. Bob Schultz, park ranger, just looked grim and sleepy, drawn into himself. Clayton Cabeen, who had been transferred to Voyageurs as its new administrative officer only three days before, and had joined us at the last minute on the spur of the moment to get his first look at the place, was excited and full of questions.

"Will I be warm enough in these clothes? How much water should I take? Will we be gone all day?"

The hour-long drive from headquarters east to Ash River skirted the park boundaries through bare, snowcapped fields on what was once pine forest rooted in the glacial moraine. It had been homesteaded after the logging, but farming was poor and most people had found jobs in International Falls or in the woods for their primary income. Half the distance to Ash River the road crossed the first granite ridge, and patches of spruce forest nudged toward the highway.

"Look at that mess!" I complained about highway construction that was broadening the two lanes to four. "That's the place where we picked blueberries last year!" I knew that the berries would come back after the turmoil of the work was done, and that the highway would accommodate the growing number of park visitors as well as local traffic, but I did not feel like being wise and gracious about the disruption. Deer were feeding among fallen trees and slashed branches by the wayside, looking to be in remarkably good condition for this late in winter. Ordinarily deer are scrawny and worn just before spring when the brushes and grasses green.

"That's easy food for them," Mary Lou commented. "Succulent." Her lovely Slovenian face had a drawn, worried look, and it seemed she was searching to find something good in the day. Anxious about keeping up once we began skiing? If so, she wasn't the only one.

"Yeah, those deer look good," Bob agreed. "But the herd may get smaller as the forests get older." Logging, fires, and brush, which follows both, provide feed for them.

"Wolf would then decline too," Frank said. "With fewer deer that happens. And the wolves don't pay attention to the Canadian border; trapping is allowed over there."

I had assumed that the Canadians were far ahead of us in wildlife conservation and protection of the environment, but in recent years I have learned this is not so. What I had taken as enlightened conservation policies were more the result of sparse population than of being forward-looking. The Canadians had shown little inclination to cooperate by controlling pollutant emissions in the new Atikokan power plant nearby, or in matching American parklands across the border after setting aside the original Quetico Provincial Park in 1909.

We passed the turnoff to Kabetogama Lake, four or five miles away and out of sight. Last summer we had canoed there, putting out from the Wooden Frog State Campground, a rocky hill jutting into the clean, island-studded lake. Very old Norway pine towered over our tent, and a sheltered bay afforded fine swimming. Now the lake was covered with several feet of ice, that in turn layered by about two feet of wind-compacted snow. The only sounds among the big trees would be a few chickadees and nuthatches, small gusts of wind whistling from the northwest, and the occasional booming and echoing of the lake ice as it contracted and expanded in preparation for the changing season. Last summer we had an unwanted concert from the blaring car radio of a group of swimmers drowning out the calling of the loons on the lake. The radio did not diminish the beauty of the place, only our enjoyment of it.

The Ash River turnoff, inauspicious as country roads go, took me into an area I had not visited before. As a very young girl, my wife had come here with her family to pick blueberries. Hers is an Ojibway Indian family, and during lean years at her home on Leech Lake Reservation, about a hundred miles south, they came here to camp and berry. Local stores and resorts bought the fresh blueberries by the quart. The picking was hot work, and the youngsters were expected to keep at it all day, and to carry the backpack crates several miles from the berry patches along an old railroad grade to the parked car.

"It was so hot, and we were so tired, that when we saw one of those cold, clear springs along the embankment, we'd jump in with our clothes on," she once told me. "By the time we'd come to the next spring, we'd be dry and hot again." On one such hike she and some young cousins fired their slingshots into the woods. A pebble hit a bear they did not know was there, and the angry animal came charging out of the brush seeming bigger than life to the frightened children, who took off at a run.

"I dropped my slingshot!" The cousin panted.

"Go back and get it if you want it so much," another yelled. "You're the one who hit the bear."

"I didn't mean to. I didn't know he was there."

They reached the car, and their parents, without having stopped for a cooling dip. The bear had given up the chase somewhere along the line—they had not bothered to look back. My wife didn't remember how much of the day's harvest had been spilled on the way, but she recalled that she still had the packing crate on her back. "I wouldn't have dared come without it," she said. "They would have sent me back for it."

Long ago her uncles and other forebears had found logging jobs hereabouts. Before then some of her ancestors had lived here, trapping, hunting, harvesting berries in summer, wild rice in autumn, maple sugar in spring, as the Ojibway displaced the Crees and drove out the Sioux two hundred years ago and more.

The Ash River Trail curved and dipped, following the contours of the land over granite hills, along riverside marsh, leading toward Namakan Lake. When the big loggers came into this area less than a hundred years ago, and most of them only fifty or sixty years ago, they followed the valley for the best access across the rough terrain. Mary Lou had been researching the local history and found forty-three logging-camp locations within the park and pictures of the temporary railroad built to haul the pine from the lake to the sawmills farther to the south and east.

The road crowded near the frozen river, then led to a small cluster of buildings and cabins, several hanging over the edge of the steep riverbank.

"These are the resorts?" I asked Frank, not believing that the sad, careworn structures were the ones that had been gerrymandered out of the park boundaries on the grounds that visitors would benefit from nearby facilities, and that the resorts would benefit from the visitors.

Frank said: "Most of them are not winterized, so they have a very short tourist season in the summer. It's a marginal business. Not enough to build up capital, which is what you need to offer modern, year-round lodgings."

It was full of ironies. If the resorters could modernize, there would be plenty of business now. Without the business, they could not hope to get the money. So there the structures sat, squat, ugly, and melancholy. Compounding the ironies, a number of resort owners just outside the park boundaries have been privately approaching park officials, offering to sell their holdings.

"Why doesn't the park buy them?" I asked Frank.

"We can't. They are outside the boundaries set by the legislation. There would have to be action by Congress before we could do that."

High hopes and expectations had gone into the exclusion of the Ash River, Kabetogama, and Crane Lake resorts from the park. Like others, I had thought this was sound public land policy. Now, face to face with the tawdriness of it, I was no longer sure. Perhaps over the years . . . with small business loans. . . . Or some larger concern might buy them out, but then it could lead to another case of Muzak piped out over the wilderness. Would local zoning keep the activities of large concessionaires within reason, within ecological bounds?

We stopped along a row of mailboxes set cheek by jowl.

"We'll pick up Ingvald Stevens's mail, in case we get that far," Frank said blandly, though I caught quick, furtive glances between him, Mary Lou, and Bob. They casually explained that Mr. Stevens was an old-timer, one of two living within the park the year around. As private holdings were being purchased by the Park Service, owners had the option of staying on in their homes or leaving. Most left but ninety-two-year-old Mr. Stevens chose to live out his years in the cabin on an island in Namakan Lake, skiing four miles one way to get his mail, chopping his firewood.

"He's a wonderful old man," Mary Lou said, "I've been interviewing him for his recollections. It's part of our oral-history program."

We packed our lunches and canteens and put on our skis. I wore heavy, corduroy knickers and thick wool kneesocks, the others wool pants, though Bob was lacing on some bilious green nylon puttees that slipped over the top of his boots and over his pants legs. I thought they looked old-fashioned and blinked at the iridescent green. We headed down the riverbank to follow Ash River out into the park.

Used to downhill skis, I braced myself for a quick slide down and a turn at the bottom. But the slide was slow and at the bottom the skis refused to turn. I'd have to learn an entirely new technique this day.

We plodded along in single file. Tiny streaks of snow meandered over the bindings of the skis, across my shoes, and I no longer thought of Bob's puttees as anachronistic footwear: they would keep the snow out of his shoes and feet, mine would be wet before long! I considered offering to trade one of my sandwiches for the use of the puttees, or maybe just one of them, but decided against it.

With a hellish roar two snowmobiles caught up with us, circled us wildly on the river ice like frantic dogs chasing their tails, lunged up the embankment and down again, and disappeared amid gas and oil fumes and a decibel level approaching sonic boom.

"We'll be away from them in a few minutes," Frank reassured me. "They'll be out on the lake ice, and we'll be going overland to Hoist Bay along the railroad grade."

Frank thinks that snowmobile manufacturers, by catering to power and speed, are sacrificing durability and dependability, and also pricing the vehicles out of the market. There once were over a hundred snowmobile manufacturers, and now about a half a dozen are left, while inventories of unsold units continue to grow. For some people in the north country, snowmobiles are means of livelihood, transportation, and survival. It seems a pity their manufacture and use have been diverted into a craze for speed, and that planned obsolescence prevents people on low incomes, who could have practical use for them, from buying and maintaining them.

Up on the old railroad embankment I wondered whether this was the one my wife had hiked as a berry-picking youngster. Hot weather seemed light years away in the damp cold, and the wind had become more blustery, scudding grey clouds above our heads. Below our trail the rusted tops of ancient cars, deserted here many years ago, protruded through the ice and snow. Frank, Bob, and summer-student helpers have been clearing such debris, but some remains to be dragged out and hauled away. It amazes me that people can so wantonly despoil a place of beauty. No wonder we have to have parks, with regulations and rangers, paid for with our taxes, if we want some beauty left. Incredible.

There were some good-sized Norway pine here and there, little trees when logging was over, but formidable now at a hundred years and older. Mixed in were spruce, birch, poplar or aspen, tag alder, and occasionally cedar, balsam, and white pine. We had gone less than three miles when the brush crowded in over the embankment and we had to break trail to make headway.

We took turns breaking branches with our hands, shoving deadfalls out of the way. It was very slow going, and I wished for my trusty old snowshoes left behind in the car, or a short-handled ax. Better yet, both. We stopped to rest and consult the map. It was past noon, we were nowhere near the lake yet. Should we turn back? Why were they asking me? I hadn't been here before and had no concept of how far we had to go once we came out on Namakan, and I said so.

"I haven't been here either," Frank confessed. He and Bob had often talked about this old railroad grade as a potential cross-country ski trail, and my visit was an opportunity to look it over. "We'll brush it out next summer," he promised. "Now that I've seen how overgrown it is."

I was reluctant to turn back before reaching the best part of the trip.

"We've only got another mile of this at most," I pleaded. "If we come out on the lake in an hour, we ought to be able to make the loop around to the next bay and go back up Ash River from there."

"All right," Frank agreed. "And if it gets too late, we can skip going all the way around past Stevens's place, and take the mail back to the mailbox."

Mary Lou still looked worried; she had kept up through the brush, though once or twice she had fallen, as had each of us. The snow was too deep to do without the skis, and the brush too thick for easy progress. Bob and I were breaking trail, certain we were nearing the lake, when Mary Lou fell again, and this time had to take a considerable rest before continuing.

"Maybe we ought to take the shortest way back," I suggested.

"She really wants to go on," Frank said. "Let's see how it goes."

The brush thinned, and on an open stretch of grade beginning the downhill slope to the lake we crossed otter tracks in the snow. Then we were in a clearing among resort cabins now used by the Park Service as a summer work camp. We took another look at the map. Stevens's cabin was still several miles away. If we omitted the side trip to his place, we would get back to our cars by late afternoon. If we went all the way, it would be dusk or later. It seemed unneighborly to me not to visit the old man living by himself and do him the courtesy of dropping off his accumulated mail. But I was also concerned about Mary Lou, who did not look at all well.

"If we take the short loop, we'll be turning left about a mile and a half up the bay," I said. "We've got to go that way anyhow, it will be faster than going back through the brush. Maybe two of us can go all the way to Stevens's, and the other two can take Mary Lou back to the car."

We ate quickly in the shelter of one of the buildings, our skis stuck in the deep snow while we stamped our feet and tried not to let on that we were cold. Then we followed the slope of the land down to the lake ice, over the frost heaves of the ice, and out onto Namakan Lake. Trestle timbers weathering and rotting dotted out into the lake, protruding above the ice in uneven, serried patterns, and then stopped. This was where the logging railroad had ended, scooping up the great pines from the log booms. The shoreline was granite topped by new forest, the dramatic, beautiful north country looking as it did when the voyageurs were here, and the fur traders, and my wife's ancestors.

The narrow skis slipped, scudded over the uneven snow, occasionally pressing down an inch or two where the wind was whipping flakes into tiny dunes. I was becoming used to the feel of the skis, getting some slide and rhythm. We passed the turnoff to Moose Bay, which was the shortcut back, and everyone went on, Mary Lou deciding that even if she brought up the rear, she would come nevertheless. A mile, another, then around the headland with clusters of granite islands—I caught a faint whiff of smoke in the numbing wind.

"He's burning popple," I said. Frank grinned broadly. We could not see the cabin yet. Then the log outbuildings came into view, soon after the cabin with windblown white smoke coming from the chimney.

We tramped up the path to the house, saw the old man through the window as he sat at a desk, writing. He did not see or hear us, and we had to knock hard several times before he came to let us in.

"I didn't expect you anymore!" he said. "It got so late. . . . I called into town and they said you were coming, but when it got so late. . . . " There were tears in his eyes. Old age? Emotion? He was a thin, ramrod-backed man with white hair and bright blue eyes. There was stubble on his face and he looked emaciated, sallow. I didn't know he'd been expecting us.

Mary Lou hugged him, looked at him closely, hugged him again.

"Are you all right?" she asked.

"Better. I've been sick three days, couldn't hold down any food. I couldn't stand up or walk, I had to crawl out to the shed to get firewood. That's why I called. I thought I had a heart attack. But I'm all right now, just weak."

She obviously cared deeply about Ingvald Stevens. It went far beyond obtaining interesting interviews for the oral-history collection. The proud, self-sufficient recluse who declined all offers of assistance had been helpless and telephoned her, and that was why she had come, pushing herself beyond fatigue, afraid we might be too late. Even when he was very ill and uncomfortable, Mr. Stevens had not accepted the suggestion that a resort operator be sent out on a snowmobile to check on him. But if Mary Lou could come. . . .

"I was so relieved when we smelled the smoke," Frank told me on the way back in. "That's when I knew he was still alive. I expected to find him dead."

"Why didn't you tell me?" I asked.

"We didn't know if we could get all the way to his place, and we didn't want to worry you."

We hauled firewood, did chores, and made sure he was able to care for himself. A snowmobile was sent from an Ash River resort to bring Mary Lou back in. The rest of us skied. It was sunset as we plodded over the lake, getting colder and windier, and I pulled the ski goggles out of my jacket pocket to shield my eyes from the sandy, driven snow grains blowing almost horizontally across the ice. The sunset colors filtering through the clouds tinted the landscape and some of the overcast mauve, light gold, and beige.

As we had left Ingvald Stevens I thought the winter must seem long to him. It does to all of us, despite our year-round love of the north. Even a self-reliant, self-disciplined man who has chosen to live by himself, maintaining a meticulous notebook and diaries, and going on skis for his mail, must relish the coming of yet another spring. So at the door I had said: "I saw a bald eagle today. They're back. He was sitting in a treetop. Very close."

Back in the protection of the park.

"A bald eagle! That's good! It'll be spring soon when the bald eagles return," Ingvald Stevens had said, his eyes no longer watery, but bright and smiling as he saw us off.

WINTER TRIP

Lynn Maria Laitala

"*Setä Antti on kuollut. Voitko te tulla?*" Cousin Laura's call woke me. So the old man finally died. I hated to take time off work for a funeral, but I'd do it for Laura.

"I'll leave early tomorrow."

We started from the Cities at 4:30 A.M. and we got to Vermilion by ten. I brought Emily along. Uncle Antti would be sent off by all the kin he had.

First we stopped in at the farm to see Mother. The dazzling white world blinded me when I stepped from the car. Clear air, clean snow. In the bright sunshine the air felt much warmer than the official ten below we'd heard on the radio. I smelled wood smoke.

Mother and the scent of cardamom welcomed us into the warm kitchen. Finnish biscuit baked in the wood stove.

She fussed and worried over us, son and granddaughter. I looked tired. Emily was too thin. It was such a long trip. She set out wild strawberry jam and took the biscuit from the oven.

"How do you manage here by yourself, *Äiti?*" I asked her.

"It's all right. Johnny comes to cut wood—and I could always hire someone. Laura brings my groceries and takes me to the doctor. I set up my loom in the parlor. Come and see."

Tidy piles of rags lay about. On the loom, Mother was finishing a brightly patterned rug of reds and blues.

Emily chattered in Finnish while we had our biscuit and coffee. She seldom spoke it at home with Ida and me. She'd learned it when she lived here as a young child.

"We should go to see about Antti," I said. Emily's eyes questioned me. She wanted to stay with her grandmother.

No, Emily. You better come along with me. We'll be back by suppertime. Should I bring groceries, *Äiti?*"

"No, that's all right. I have peas soaking for soup."

We traveled on to Sawyer. Laura's ancient pickup truck was parked in front of her house. I walked in without knocking— old habit—and startled Laura at her desk. It had been over a year since I'd seen her. There were new lines in her face—from weather, I thought, not worry. She rose to hug Emily, then me.

"How good to see you both. Can I get you some coffee and some *korppuja?*"

"No, thanks. We came from *Äiti's.* When is the funeral?"

"Tomorrow. But Jussi doesn't know yet. He's caretaking up at Skidway Lodge. I thought we could drive up to Basswood now and get him."

I looked out the window at her truck. "Let's take my car," I said.

Laura laughed. "There's nothing wrong with my truck. Johnny's been keeping it fixed up."

"Where is he?"

"He's around. He's been staying here."

"Does he go to school?"

"No."

I dropped the subject.

"Let me make a thermos of cocoa and some sandwiches and then we can go." She went into the kitchen. Someone came in the back door.

"Johnny!" Emily ran back to see him. I followed more slowly.

"Hi, Dad," he said. Emily was hanging on his neck. "Thought I'd run up to Basswood with you."

"How are you doing, John?"

Johnny sat down and looked at the floor.

"I'm fine," he said.

"Load up," Laura ordered. "I'm ready." She looked us over. "Is that all you people have to wear?" She rummaged in the closet and threw out snowpants, hats, scarves, and mittens. Emily bundled herself up. I added a scarf. Johnny stayed bareheaded and barehanded. Laura made another attempt. "It's only ten below but the wind might be bad on the lake." Johnny grinned and stuck the tips of his fingers in the pockets of his jeans.

We took my car. The new county boat landing was built on Frank Silvola's place. Had they burned down his cabin? The road ran right off the landing on to the ice.

"Ice is two feet thick, Andy," Laura said. She'd sensed my hesitation at the shore. The road was much smoother over the ice. Plowed clean, perfectly level.

Emily bounced around on the back seat.

"Driving on the ice!" she said. Laura looked at me. We'd lived our winters on the lake. How could things be so different for my daughter?

The road ran down the middle of Wabeno Lake, curved away from the falls, and disappeared into the woods at Four Mile. At the end of the portage smoke curled from the little store.

"It's just a winter caretaker, but he'll sell cigarettes to the desperate," Laura said.

"There's a huge locomotive wheel right off the landing there, under the ice," Johnny told Emily.

"Logging train," I said. Johnny knew.

The road dropped to the ice again. I still knew Basswood Lake as well as anything in this world. The road branched out occasionally to this resort or that.

"This will be the last year for most of them, Andy. I think only Hubachek will be left after '61."

"I know." The government had banned the resorts. Some would be moved, others burned. I asked Laura about their fates. Beautiful buildings — two stories of white pine logs — had already been burned. I felt sick at the waste.

Jussi's little caretaker cabin smelled of fried pancakes, coffee, and the damp wool socks steaming dry by the stove. Laura gave Jussi the news.

"Antti dead!" He sat down. We all sat. "Well, he was sick now, for awhile. But dead." He rose and set out heavy chipped mugs, coffee, and a can of condensed milk. "He saved his suit for the funeral. It's at his shack. I know where."

"You were together a long time," Laura said.

"I was in his first logging camp, just after he came over from Finland. Yah, we partnered off and on for almost fifty years. Logged, trapped together. That's all done now. All done."

"What will you do next year, Jussi?" I asked. Next year Skidway would be ashes and pine seedlings.

"Time to hang my teeth on the nail."

"Antti wanted you to have his cabin," Laura said.

"Nah, I'll stay in town." Jussi picked up the mugs and put them in the dish pan. "I gotta fix up some things here, then I'll come with you guys, show you the suit."

"Let me take Emily out driving on the ice," Johnny asked.

"Please, Daddy?"

I considered whether twelve-year-old Emily could control her brother.

"Okay."

Laura and I walked up to the little point of rock overlooking the great frozen lake. Five hundred miles around if you paddled next to the shoreline, half of it in Canada.

"Who will see it like this anymore?" I asked Laura. I remembered how we would snowshoe from Sawyer on Friday night, often as far as the Canadian ranger station, and go back home on Sunday. Fifteen or twenty miles each way, in all kinds of weather. For the glory of the frozen world.

"People will always find a way," Laura said.

"What were there, twenty resorts? Twenty-five? They could have just banned new building and left these until their owners retired." How welcome was that first scent of smoke and glimmer of lamplight in a far-off window on those long treks from town.

"That isn't the way they do things, Andy."

Far below us on the ice Johnny was whipping my car in circles. We watched as the car came to a stop and Emily and Johnny walked around it, switching places. Emily's first time at the wheel. There wasn't a car in sight. Little danger of a collision. She'd have to run it up on shore to do any damage. The car started in jerks, then traveled slowly along the ice road. Emily kept carefully to the right.

"Are you ever coming back up here, Andy?" Laura asked. I knew she meant move back home.

"It's too late, Laura."

Below us, Jussi opened his cabin door and threw cold pancakes to the Canada jays.

"He always cooks extra," Laura said.

We started back to the shack. Jussi was putting on his coat.

"I'll show you Antti's suit now."

We hailed the children back from the dock. Emily drove up sedately, precisely, and stepped from the driver's side grinning. She and Johnny climbed into the back seat with Jussi. Old Jussi was melancholy.

"Used to be life on this lake," Jussi said. "There were some characters in those logging camps. Yah, and there were still Indians when we got here. Antti partnered up with that Charlie Boucher for years. He almost married Charlie's sister, too. Used to be three, no, four villages on this part of the lake. Used to trade furs with the Canadians right here, but before my time. Government took all the Indians. Now it takes everybody. Antti's better off, dead."

No one said a word on the rest of the trip to Antti's cabin. On Wabeno Lake we had to jump the snowbank but the snow wasn't deep and the car drove through it easily.

It was colder inside Antti's cabin than it was outside in the sunshine. Jussi went to the bed and pulled a torn cardboard suitcase from underneath. We all stood and watched expectantly. Inside was a heavy wool suit, once black, now tinged with green.

"Last time he wore this suit he danced on the boat from Liverpool," Jussi muttered. He lifted it from the suitcase.

"Look!" cried Emily. Three little books—bankbooks— had dropped from the folds of the suit. Emily picked them up. Three different banks.

"Look at this. Look at all the money he had," she said. I took the bank books from her. Frequent and regular deposits were listed from 1912 through the 1920s in all three books. Apparently Antti hadn't trusted banks well enough to put all his money in one account.

"Lots of beaver pelts," said Jussi, looking at the figures. All the entries stopped in 1929. I totaled the figures in my head. Antti had lost more than three thousand dollars. It was a lot of money for an immigrant worker in 1929.

"He used to talk about going back to Finland to find a wife," Jussi said. "He almost made it."

"Can't we get out the money, Daddy?"

"No, Emily. The banks failed a long time ago. There is no money."

"What happened to Uncle Antti's money? Who got it?"

"It was the Depression, honey. No one got the money. It was just gone."

We took the suit. Johnny pocketed the worthless bankbooks. We looked around at the few dishes and pots, then went up to the woodshed. An old kick-sled lay on its side, a half-cord of split wood spilled from its pile, and rusty traps hung on the walls.

"Anything you want, Jussi?" Laura asked.

"Nah."

"You can stay at my house overnight," Laura told him. "I can take you home after the funeral."

"Nah. Take me back now. Dead is dead. What's a funeral?" I followed him back into the cabin. He looked around, then picked up a mug from the shelf—an ordinary coffee mug with one line painted near the rim. "I'll take this."

We drove Jussi back to Skidway. He got out of the car without looking back. I shut off the motor and watched him climb the hill to his cabin, shoulders stooped under torn mackinaw, coffee mug dangling from bony fingers. I had never seen him look so small or lonely.

The trees along the lake shore were silhouetted against a red-streaked sky as we headed back to Sawyer. Two trips to Basswood in one day. I wondered over it to Laura.

"Do you remember the time we snowshoed home in the wind, wrapped in blankets from the ranger station? It took us fourteen hours. I've never been so cold."

"I thought we were dead," Laura said. "Can I make you some supper?"

"No, Äiti's expecting us. Want to come to the farm for some pea soup, John?"

"No thanks, Dad. You can let me off here."

I dropped him off on the road to Vermilion.

"Will you be at the funeral, Johnny?" Laura called out her window.

"Sure. One o'clock, right?"

"Right."

He stood, hands in pockets, back hunched against the wind, waiting for a ride to town.

I pulled up behind Laura's battered pickup. I wanted to talk to Laura, but Emily understood Finnish so we had no privacy.

"I'll see you at the funeral, then," Laura said, hauling out the shabby suitcase with its ancient suit. She closed the car door, then opened it again.

"Antti had a good life. He was a free man. Don't grieve."

I said nothing. All I felt was grief. Emily spoke.

"He had a good life, Täti Laura. It was the best."

The air had grown bitterly cold. Emily and I drove home to Äiti's warm kitchen.

GOIN' TO THE DOGS

Judith Niemi

It was the middle of a January night out in the woods up north of Two Harbors. A lot of us were there, stomping our feet, huddled around bonfires of scrap lumber. Forty below zero—that's forty honest degrees, none of this windchill stuff—and we had nothing to do but search for our friends among identical woolly lumps of humanity and wait for dog teams to trot out of the woods.

"Let me get this straight." A Beargrease rookie was with us, mystified by northern Minnesotans' idea of winter pleasures. Her voice came trailing thin out of her hood. "We're going to stand here in the snow, like my vision of Siberia, and when some dog teams come in we'll maybe get to pat them a little? *If* they're friendly."

"And talk with the mushers," I say. "If they aren't too busy."

"OK. And then we're going to drive fifty miles up the

road to another cold place and wait around a few hours to do it again?"

It isn't easy to explain, this fascination with the John Beargrease Sled Dog Marathon. You have to be there—Lake Superior slapping ice onto the cobble beaches, black spruce trees shining with hoar frost. The waterfalls run under fluted columns of ice, and every year my friends and I promise ourselves that we'll take time out from the race to snowshoe or ski at the Temperance River, to soak in the beauty. But always we're seduced into chasing after dogs. It's an annual defiance of winter, a celebration of old ways. Forget rationality. This is ritual.

The Beargrease, an annual round-trip race from Duluth to Grand Portage and back, evolved from a few North Shore mushers competing in a race they called the Gunflint Mail Run during the late 1970s. Originally a one-way race from Grand Marais to Duluth, the Beargrease took its name from John Beargrease, a member of the Grand Portage band of Ojibway. Between 1887 and 1899 he carried the mail by boat and dogsled along Lake Superior's North Shore, a one-man, north-woods version of the Pony Express, undaunted by wolves, blizzards, or the wild terrain.

Each Beargrease racer carries a commemorative mail pouch on his or her sled, along with traditional and essential safety gear: knife, ax, sleeping bag, and snowshoes. The country along the trail is still mostly uninhabited, and checkpoints are thirty to sixty miles apart. A musher who has an accident or loses a team could have a long, lonely walk out.

Witnessing the annual event is part of a North Shore winter. Hardy residents volunteer to operate ham radios, guard road crossings, provide food, and control parking. Some tourists and skiers stop by, lured by curiosity. And then there are the aficionados, the mushers' friends and families, and the growing number of sled-dog groupies who leave warm homes and take time off from respectable jobs to follow the race up and down the North Shore.

At almost five-hundred miles, the Beargrease is one of the longest sled-dog races in the Lower 48, a challenging course. Most of the race happens out of sight, up in the rugged hills over Lake Superior. But unlike other long races, it has numerous checkpoints and unmarked trail crossings where an active spectator can share with the mushers the exhilaration, the cold, the sleep deprivation.

The race begins at Ordean Junior High School in Duluth. This is the wildest single part of the race. Traffic is backed up for blocks, the bleachers are packed, and teams bolt out of the starting gate every two minutes. (The staggered starting times are adjusted later, as mushers take their breaks.) No one, by the way, every yells, "Mush, you huskies!" The word is "Hike!" or simply "Let's go," but you can't hear it anyway. The dogs waiting and the dogs in the gate are barking and leaping, and every veteran dog among them knows how to count backward with the starter. A millisecond before the starting signal, the dogs take off, the sweating handlers letting go just in time not to be dragged up the hill.

Spectators new to distance racing are often surprised at how small the dogs are. Odd-looking, too. Some are classic silver huskies with remote ice-blue eyes, but other dogs look scruffy, like nothing the American Kennel Club would sanction. You see signs of coonhound, whippet, German shepherd, or black lab in these beasts. One year there was even a team with a lot of standard poodles, to the delight of the working press and the disgust of some other mushers, who complained that the poodles had no trail manners. Poodles aren't born with pack instinct or a passion to run; this rough bunch had been brought up from puppyhood in husky culture.

Whether pure Siberian or Alaskan or some other breed, the dogs share an aristocracy of nature. Bred and trained for stamina, speed, and the sheer will to run, they are the Olympians of their sport. Watching them tear down the trail, you think that just being alive is a finer and more exciting thing than you'd suspected.

Many fans skip the starting gate and watch the dogs farther up the trail. As they cross Skyline Parkway, the dogs are quiet, pulling hard; the mushers are running too, lightening the load on the hills. Ten miles away, the crowd at the Idle Hour tavern cheers them along. Miles farther, back in the woods, a few women sitting around a campfire by the trail offer hot chocolate to any musher who wants to stop.

Most teams don't rest until Two Harbors, in the middle of the night. Even then they are too excited to linger. One year— what a great time that was—we of the loyal Two Harbors audience got recruited by trail judges to help with traffic control. Teams of eighteen to twenty psyched-up dogs were flying down the trail right on each other's tracks and colliding where a hairpin turn and a narrow bridge met. We formed a human chain; mushers shouted in French and English. "I don't believe this," said a small voice in the blackness. "I haven't even been introduced to these dogs, and somehow I'm gonna be grabbing them and asking them please to stop?" But these agreeable dogs didn't mind, and things got pretty much under control. One team refused to cross the bridge and dragged their musher through the open creek. Another team split their vote on the bridge. It's a contact sport, definitely, but few serious injuries happen.

By 3:00 A.M. all of the fastest teams had passed, and we went off to bed. It was plenty cold in the outhouse at our motel, but the management had thoughtfully provided a warm styrofoam seat. Out the open door I watched the slow, icy waves of Superior, the glittering stars, Orion stalking westward. The mushers, I thought, shivering, are going to be running all night.

The nicest part of the race is the quiet woods far up the shore. College students controlling traffic at one back-road crossing build snow caves to sleep in. A few people ski on the dog trail near the Temperance River. It's always peaceful here, sun filtering through the trees or quiet snow falling. A hundred or more dogs are asleep in the hay and snow. Many mushers prefer to rest their dogs during the day, when it's warmer, and to run at night. All teams are required to take one twelve-hour stop, and most do it well before the half-way point.

There are no play-by-play broadcasts on the Beargrease trail, but race news travels quickly, spreading through the trailside crowd before it hits the media. One year the big story going around Sawbill was that Joe Garnie's team had collided with a bridge, and the sled brake got smashed. Joe drove forty-some miles that night to the Finland checkpoint—on a rough trail, with no brake, with a full team of twenty powerful dogs. He'd improvised a kind of drag out of a duffel bag and reportedly was asleep at Finland.

Along the roadside a few craftspeople sell parkas, mitts, or nylon booties for dogs. High chic at the Beargrease is enormous beaver-fur mitts, mukluks, fur ruffs, voyageur sashes. Stylish wear is snowmobile suits, Sorels boot pacs, and knit caps with sled dogs stitched on the front. Word goes around that a local woman is giving away doughnuts fried in real bear grease. The doughnuts are light and delicious; the coffee, too, is free.

Around campfires the mushers' assistants brew up great kettles of mink meat, beaver, and the sponsor's brand of dog chow. Tiptoeing around sleeping dogs, we chat with some of the crew and with the mushers, when they aren't catching quick naps or tending to the dogs. The mushers are mostly an engaging bunch, men and women who forsake many of life's possibilities to spend their days as the trainers, coaches, cheerleaders, and fellow athletes of these magnificent dogs. Mushing isn't a hobby or a vocation, it's an obsession. Most of the drivers breed, raise, and train their own dogs, and they live with sixty or a hundred or more of them at a time. All for the pure joy of speeding through the winter woods behind a fast team and for the perfect moments of uncanny communication with their perceptive lead dogs.

Prize money is good in some races. The Beargrease now has enough commercial sponsors to divide $50,000 among the top ten finishers. But only a few mushers expect to meet their expenses, even in a good season. The impractical, unorganized aspect of the sport is appealing. No franchises, no leagues. If any betting action is going on, beyond the casual dares between crews, I haven't found it yet.

You invest yourself in the race by picking a favorite musher or two to follow, be they front-runners or simply people you like. Last January we knew just where to find Jamie Nelson, the musher from Togo, Minnesota, who finished first in 1988; it's not a quiet rest spot, but her dogs have stopped there for several years, and they aren't about to change their ways. One of our friends went off to find Ray Gordon from Wyoming. He's always short of handlers, and soon she had signed on and was taking an ax to a disgusting-looking block of frozen chicken parts. Another musher had just agreed to take a TV camera on the sled for the next leg of the race. Individualists that they are, mushers try to cooperate with reporters and explain their sport.

As the dogs wake and stretch and get harnessed up, you get a chance to collect data on the key players. Bonnie slept on hay instead of straw and had to drop out early with asthma. Luke raided the food stash and got almost too bloated to run. He was mortified at being demoted to wheel position for a while. Even though I never collected baseball cards as a kid and was only a lukewarm spectator of hockey and football, I start imagining a line of sled-dog cards. My prize would be a card of Trace, the small black Alaskan with the foxy face. What attitude that little bitch has, biding her time back in the pack, knowing her spirit is going to spark her team at the end.

Two days down the trail, more or less, the teams reach Grand Portage. A pow-wow is held at the hotel honoring John Beargrease, who once lived there. The first musher in gets a prize of beautiful beadwork. Along the lantern-lit trail where the teams come in, a company of voyageurs in late eighteenth-century dress has pitched a tent in the snow. Someone is playing a fiddle at the campfire.

We went cross-country skiing here last year, along trails well groomed but punched full of deep moose tracks. When our ski trail connected with the race trail, we couldn't resist and were drawn back to the dogs. You just have to keep checking back over your shoulder, listening for the soft swish of runners or the puffing of dogs' breath and being prepared to scramble out of the way.

Many of the mushers take a good rest at Grand Portage, sleeping in rented mobile homes or at the Radisson Hotel. With so many teams collected in one place, there's more news and dog gossip around. As the teams begin pulling out, the long layovers over, it's finally possible to see who is leading and what strategies are developing. Some crews say nothing to spectators and reporters. Some will tell anything but, of course, they could be making it up. Don't fall for the psyche-out, all the mushers will say. Just run your own race.

Warm weather is worse than cold weather, and new snow is difficult. "I tried a new dog at lead," said one musher. "He's the only one who'd really go in those drifts. He and I were just disgusted with the rest of the bunch." Dogs need to rest longer if they aren't genetically blessed with abrasion-resistant feet. Some dogs like to chase, so their drivers will let another team go ahead.

The handlers also talk runners' diets. One handler swears by lamb chops. A lot of Minnesota dogs like venison. Last year Dee Dee Jonrowe's Alaskan dogs were snacking on salmon, courtesy of her sponsor, a fishery.

Checkpoint by checkpoint, the teams are getting smaller. The mushers and a veterinarian examine the teams at each stop, and tired dogs or dogs with sore feet are retired to their boxes. Sometimes out on the trail a dog becomes ill or too tired, and it comes in riding in the basket of the sled. No new dogs may be added. The smaller teams have less strength, but they aren't necessarily slower. Most mushers will have only a small number of their most seasoned dogs cross the finish line.

The mushers also watch out for their dogs' morale. "The dogs really have to count on you to pace them, to know their limits," says Cynthia Dech, a retired sprint racer and enthusiastic sled-dog fan. "If you lose their confidence, you've had it." Besides ten happy years with dogs, Cynthia has had a long career as a manager in the computer industry. "Absolutely everything I know about managing came from running the dogs," she says.

This last part of the race always seems sort of foggy to me. There are too many teams to care about, scattered too far. We shift into four-wheel drive and tear around the back roads, but at all the crossings we find little dog prints already headed south. On the return route the trail makes a jog to touch the lake shore at Tofte: 500 vertical feet down, then 500 feet back up to the top of the old Sawtooth Mountains. It must be along here that some mushers and dogs hit the wall, or whatever they do. What does a musher hallucinate? Do dogs hallucinate? Are we still having fun? Does it matter?

Memories of several races are mixed up in my head. A car in the ditch. Ours. Losing track of our carpool. Hanging on to a dog collar at 4:00 A.M. so an athlete named Mac wouldn't steal Cisco's food.

I vividly remember a musher sitting by his campfire, sadly watching his exhausted dogs sleep curled together in the hay. "I blew it," he said. "I ran them too hard. As long as I could see the front-runners at checkpoints, I thought I could catch them. From here, I'd just like to finish. At my own pace."

At Beaver Bay the mushers still left in the race trudge up the birch-covered hill to pay respects at the grave of John Bear-grease; they follow the Ojibway custom of offering tobacco. Each year only about half of the starting teams will finish the race. For some, to scratch means defeat. Others are content just to participate. As a musher who had come all the way from Montana explained: "To be really competitive in this race, I'd need to build a whole different team. But these dogs are my friends. Hey, we gave it a good run."

Mercifully, there's one last, long pause, a mandatory four-hour rest for each team at Two Harbors before they start the final run to Duluth. May I never have to explain to a dog team what ambition of mine is important enough that, having raced more than 420 miles, they should get out of the straw and back on the trail. Sometimes there's a little grumbling and a strike or two. But they go. With luck, there's even a little bark, canine version of a high-five.

Down in Duluth, the spectators make a big weekend of it while they wait for the racers to return. Dogsled rides, a cutter and buggy parade, a cute-puppy contest, and mutt races. The hope is that the winners will cross the finish line at prime time, to noise and crowds and prizes from sponsors. Sometimes it's a neck-and-neck finish, with teams passing each other as they fly down the hills into Duluth. Often as not, however, the first finishers come in during the small hours, with only a handful of reporters and diehard fans still awake.

After the first teams cross the finish line, other teams are strung out along the back trail, hours behind. Some of the mushers come in too exhausted to be coherent. Some are keyed up and can stand around talking for hours.

Joe Redington, Sr., was one of the finish-line crowd's favorites a few years back. At sixty-nine he was the oldest racer and was known as "the father of the Iditarod," the Alaskan race that helped popularize sled dogs again. Joe had been penalized an extra eight-hour layover for missing a checkpoint; well rested, he sped down the track, passing mushers forty years younger. He arrived seventh, joking about his bad habit of stargazing, and his wrinkled face had the look of a man contented with life.

Julie was another crowd favorite. When Dave Oleson, the tall boyish musher from Ely, came down the trail, he had only four dogs in harness; five are needed to qualify. Twenty yards out of the finish he stopped. The fifth dog was riding in the sled basket, looking a little embarrassed. Dave and the little white dog gazed at each other a long time. Then he gently lifted her into the harness.

"What's the dog's name?" someone called.

"Julie."

"Go, Julie!" the crowd yelled. She did.

There's no climactic ending to the Beargrease ritual. At some point you just notice that you are sated. No wish to watch even one more team come in. No more thoughts about buying a husky, just one. You notice a doggy smell when you turn on the heater in the car.

So, home again. But since the Beargrease racers have thirty-six hours after the winner to qualify as finishers, you send your thoughts for a moment off to the Northeast—to Finland, Beaver Bay, Fox Farm Road. Men and women are still out there, carrying their mailbags, talking to their dogs. Hope they make it.

A TALL MOOSE
COMES CALLING

Ted Hall

A tall moose came out of the woods the other morning to see what was going on around here and everything that went on around here for a while after that was connected with the moose.

Harry Oveson appears to have been the first person to know there was a moose in the neighborhood, although he first thought it was a large bird flying past the east windows of his house. Harry was seated at his kitchen table enjoying his first cup of coffee of the morning, but he is interested in birds and so he arose and went to the window to see what kind of bird it was.

That's when he saw the moose, and when he remembered that the bottom sills of his kitchen windows are six feet above the ground and that he had seen the moose's head pass across the top half of the windows, he realized that he was looking at a very tall moose. He calculated that it was eight feet high, and he was looking at it from a distance of twenty feet, and he recalls thinking, "What in hell . . . ?" and concluding, "If I'd had a bad heart, I'd have keeled right over."

Mr. Oveson picked up his camera and headed for the back door on a collision course with the tall moose. Then second thoughts intervened and he put the camera down and picked up a .22-caliber pistol because it occurred to him that the reason that tall moose was in his yard was probably because it was being chased by dogs and his thought was to intercede on behalf of the moose.

The moose crossed Harry Oveson's yard on a northeast-southwest diagonal course heading southwest, paused in a cluster of four birch trees, and stood looking out over the ice in the old gravel pit that is locally known as Oveson's minnow pond because it was once used to raise minnows in until it was discovered that minnows take a long time to grow and it's better business to go out and catch grown-up minnows in the lakes than to wait for them to grow in a place you have to pay taxes on.

It struck Harry Oveson that the moose had neither run nor sauntered across his lawn, but moved at a speed somewhere in between, and, because the ground was frozen, had left no footprints.

The moose now pointed north and went behind Vernon Oveson's house next door and disappeared for the moment from Harry Oveson's view. Harry Oveson says he got another glimpse of the moose, but it was just a glimpse of the assend going toward the center of the village down the road.

Just about the time Harry Oveson was firing up his coffee pot for that first cup of coffee, Russell Hanover was setting forth on his morning job accompanied by his year-old bulldog, Ted. Usually Mr. Hanover makes a three-mile swing that takes him up to Point O'Pines and back and he varies his route by going into side roads that lead him along the lake. This turned out to be a good day for jogging and he swung into Birch Point and then headed for home. He came out the Birch Point road and turned west on the Ranier road and was immediately aware that something was blocking the road about two hundred feet ahead. He fixed the time as about twenty minutes to seven because he times his morning job so as to be back home and ready for the seven o'clock news, and what he saw blocking the road was a moose standing in the road and facing south.

To Mr. Hanover's relief, his dog Ted did not see the moose or at least did not admit to seeing it. As Mr. Hanover approached, the moose proceeded into Ed Baron's yard, and when Mr. Hanover got there Mr. Baron was out in his yard complaining that something was amiss with his camera. Mr. Hanover said he told Mr. Baron that he didn't know anything about moose but he did know something about cameras and he investigated and discovered that the reason the camera wasn't working was because it had been loaded with brittle old film.

Mr. Hanover said his first thought when he saw the moose there was that it was there because it was sick. "But it appeared to have a full deck," he said and it ambled slowly away.

Now we take you to the home of banker Jay Nims, who, like Harry Oveson, was having his first cup of coffee and was seated in his pajamas in his front room looking out across the lake. Suddenly a moose came from the east, he says, and passed about fifty feet from his window on its way to the west. It had no antlers and didn't move at a fast pace. It paused at his four-foot snow fence, then bounded over it and went toward the home of athlete Bronko Nagurski while Mr. Nims went to get his camera. In the seventeen years he has lived there, Mr. Nims says, he has seen deer and bear and just a year before saw a deer on the ice out front with a pack of dogs after it.

Over at the Bronko Nagurski home Bronko Nagurski had gotten up first and made the breakfast coffee and had moved into the front room to read the morning paper. His wife, Eileen, was still at the breakfast table when she caught sight of movement between a spruce tree and a cedar in Jay Nims's yard next door. That is a distance of about one hundred feet.

Mrs. Nagurski saw that she was seeing a moose and she called out the news to her husband. Their son Kevin, who is seventeen years old, had entered the front room in his pajamas with about half an hour to get dressed in time to catch the school bus. When Kevin heard about the moose he ran out the door, looking for the moose and calling for his camera and some film. Mrs. Nagurski said she didn't know where Kevin's camera was, let alone film, and so she went out the back door and discovered to her surprise that the moose had changed course and was cutting across the driveway and getting closer to her. She remained standing in the doorway and the moose passed about twenty feet from her and towered above the parked car that separated them. Mrs. Nagurski says that the moose did not seem to see her and walked across through a grove of eight large spruce trees and into the Henry Erickson place, when the barking of Rascal, the Nagurskis' dog, caused it to change plans and circle back and head toward the road. Rascal stayed a discreet distance behind the moose and stopped following it when he reached the Nagurskis' roadside garden.

Bronko Nagurski says he got just one glimpse of the moose right after his wife called his attention to it coming out of Jay Nims's yard, and then he went into the bathroom to shave and that's how he missed another look at the moose as it passed Mrs. Nagurski when she was standing in the back doorway.

The hunch is that this was about the time Russell Hanover and his bulldog Ted were emerging from the Birch Point road, and the trail of the moose seemed to evaporate there.

Mrs. Nagurski said it was the second moose she'd seen in her lifetime—the first one crossed the highway east of Atikokan as she was motoring to Thunder Bay.

Russell Hanover said it was the first moose that he'd seen and that this was ironic because his father, the late Dr. Ralph Hanover, had traveled all over the county on his rounds and never seen a moose. "—and here is his kid, out playing, and sees a moose."

Harry Oveson said that up at his fish camp on the lake he goes seven or eight years between seeing moose.

The moose news spread and there were even reports that Bronko Nagurski had gotten a picture of it. Russell Hanover stopped for a haircut on his way to work, and when he got to his drugstore his employees already had heard about the moose. Mr. Hanover says all this excitement is understandable. "It's so unique. He's really our king of beasts. Everybody is surprised but kind of glad, it's been so long since there's been a moose around here."

No one who saw the moose was able to say if it was a bull

moose or a cow moose. It had no antlers, but that's the style for bull moose that time of year.

The big question that went unanswered was: *and whatever became of the moose?*

Not long after the visiting moose disappeared along the Ranier road, Game Warden Marvin Smith got a call. At a certain residence along that road, he was told, he'd find a moose head in a wheelbarrow.

Warden Smith and Warden Tom Kjellberg pursued the tip and they were just a few hops behind the angry rumor that the sociable visiting moose had been done in. They found the address and the wheelbarrow and the moose head. But when they looked for the rest of the moose they found no trace. Then they examined the moose head in the wheelbarrow and found it had been dead long before the sociable moose livened things up along the Ranier road.

There had been a party and somebody mentioned the wandering moose and somebody had an old moose head and there you are.

And by the way, whatever became of the moose?

LOVE AND DEATH

Matthew Miltich

"*Love animals, don't eat them!*" demands the bumper sticker on a late-model car outside the Depot in downtown Duluth. I read this on an autumn evening when I've come to town to hear William P. Kinsella, author of *Shoeless Joe*, read from his fiction. Pictured in silhouette on the bumper sticker are animals that, except that we eat them, would likely not exist at all. The dusky air is chilly and I shiver slightly inside my blanket-lined denim jacket. Perhaps, I think, we love animals *because* we eat them. Let me explain.

I will not argue with those opposed to the eating of meat about issues of nutrition, for they are either right or wrong in their claims, or right about some things, wrong about others. I think though, that people who oppose the killing of animals for food are really opposed to killing, because they cannot accept it as such. If this is true, they are not different from most other Americans; they are simply more in tune with what is implied by what they eat.

In the twilight years of the twentieth century, most Americans kill the food they eat not by their own hands, but by the same means with which they are supplied most of the other essentials by which they live: industry. Killing is something they cannot or will not do because it is uncomfortable for them, or being removed from the land that gives life they've lost the habit or forgotten how. And so they've become afraid of killing, afraid of death generally, and thus afraid of life, which is bound up inseparably with death. Real love of life must embrace all the processes of life.

In the rural and still-wild places where northern Minnesotans live or that they have close and easy access to, such an embrace is still possible, even probable. Northern Minnesota gardeners, farmers, fishermen, hunters, wild-food gatherers, and all those who take life to sustain life should be lovers. What is required is a certain centering on the cherished object. The gardener must love the plant that will be her food; the farmer must love the land and the plants and animals it supports which in turn support the farmer; the hunter must love the deer; the gatherer love the wild mushroom and the wild places that give it life.

But this is not as easy today or as simple as perhaps it should be, for even in these primary and primitive pursuits, industry has interposed its technology between people and the objects of their loves and labors. The gardener becomes enamored of implements instead of plants, the farmer of machinery instead of the farm; the hunter loves his armament, the gatherer his guidebooks. When life gathering becomes centered on paraphernalia rather than the life it seeks, killing ceases to be sacrificial and becomes a kind of murder instead, and it is right that we should censure and abhor murder.

Nonetheless the terms of sacrifice are not denied to modern people. Awareness of the sacrificial relation between the life taker and the life giver is ancient but also timeless. Its central principles are respect, gratitude, and humility. An Ojibway prayer to a deer slain by a hunter proclaims, "*Without you, I hunger and grow weak. Without you, I am helpless, nothing.*" The prayer acknowledges the deer's right to life but also the need of the hunter;

it expresses both humility and gratitude; its terms are those of sacrifice, not of murder. These are precisely the terms that any people living in any time must understand and accept to live in right relation to the lives sacrificed for their sakes. We can live and cherish life for its own sake as well as our own.

If we deny the terms of sacrificial relations with the lives that feed us or allow industry to replace them with the cold transactions of the marketplace, we give up one of the important aspects of our humanity. Sacrifice is as much a matter of attitude and understanding as of necessity and action. What we must cultivate then is not just our garden, but our awareness as well.

To assume that there is no longer any value in such an understanding because industry supports us is to forget what supports industry: land. When we allow ourselves to forget the land that gives us life, we begin to regard living forests as "natural resources," to regard farms as "agribusinesses," to regard prey as "game." If we fail to come to terms with sacrificial killing, we are in danger of murdering the very land that must live to sustain our lives.

If we regard the industrial processes that destroy or obscure our awareness as benevolent because they save us sweat or discomfort, we might turn to our other sacrificial relations and ask how much we are willing to relinquish of them to be relieved of work or pain. Would we be willing to entrust childbearing, child rearing, or marriage to industry? In these relations, our own sacrifices are rewarded by attachments, and understanding and love. These rewards have not been increased or improved upon by industrial technology.

The ancient Greek civilization understood and appreciated the beauty and dignity of sacrifice and ritualized their understanding in religious ceremony, which eventually evolved into the drama of tragedy. Greek tragedy accepts both the inevitability of death and the pity which that arouses in us for one another. In this drama, the tragic hero is the sacrificial victim by which life is redeemed. In his tragic recognition, the hero realizes and accepts his flaw, is humbled, and finally summons the courage to face death. What is most remarkable about tragedy, though, is the emotion it awakens in an audience. Witnessing the tragedy, the audience is roused to pity and fear for the hero. The hero's passion is the means of awakening compassion in the audience. It was understood in that ancient society that the release of such powerful emotions in people made real personal change possible. By such a ritual encounter with death, the Greek citizen was able to confront life more openly and with newly composed courage.

Today we might also be awakened and encouraged by the tragedy of autumn. Those of us who love autumn above the other seasons might ask ourselves, why? What lends the season its special intensity, its poignance? It is nothing less than the death of the year. Its perfumes are composed of the individual deaths of countless leaves and grasses and other plants. Its hues are the colors of last struggles and death. It is the season of the harvest and the hunter. For the creatures that migrate, it is the time of greatest threat, greatest danger; their flight before the white giant, winter, is a gambit.

The season of mists and mellow fruitfulness is a time to harvest and hoard, to take and put away life to sustain life through the great death of winter. The sweet smell of put up hay is the smell of summer's life, taken in its prime to keep for the season when grass won't grow and water won't flow. The blood of the butchered steer or deer is sacred blood, spilled not wantonly, but in the sure knowledge that living flesh feeds upon dead flesh and life goes on.

If this is sad and tears must be cried, so be it. In some seasons, the price of love is tears, and love can only proceed from knowledge. Ray Bradbury wrote, "We are the creatures that know and know too much. That leaves us with such a burden again we have a choice, to laugh or cry. No other animal does either. WE do both, depending on the season and the need." We know, if we are honest, that we are killers, by deed or fiat, whether we eat meat or plants; we can choose to honor the lives that are given to us, or to ignore them.

Shelley wrote, "When Winter comes, can Spring be far behind?" In northern Minnesota the answer is, "Yes, it can!" Be that as it may, spring will come at last; a proper appreciation of autumn may be what we need to be ready when it comes.

RICING AGAIN

Jim Northrup

Luke Warmwater woke up wondering where he was. He remembered being at the party and remembered being one of the last
ones awake. It all came crashing back. He was at Mukwa's house
and he was supposed to meet that woman who was his new rice
partner.

His head hurt and he knew his breath was foul. His ribs
hurt and he wondered if he had been in a fight. No, no one had
punched him, it must have been the way he slept. The couch was
responsible for the rib ache. The alcohol was responsible for his
parched throat, queasy stomach, and shaking fingers. The headache came from the guilt and the excessive amount of alcohol he
drank.

His bloody eyes scanned the room looking for something to put out the fire in his throat and settle his stomach. He
knew the guilt feeling would go away on its own. He saw a twelve-

pack that had been overlooked last night. The twelver still had seven cold ones left. The world didn't look so bad now.

His old car was outside and the cold beer in his belly was making him forget his assorted aches. He drained the bottle as he stood outside and drained his bladder.

The canoe was still on the car, and under a quilt in the back seat was his new partner, Dolly. She looked a little tough as she snored away her share of the party.

Dolly came from the other side of the rez. He knew he wasn't related to her but had heard of her family. He also remembered her laugh from last night as she playfully punched him on the arm. He was trying to tell her how pretty she was when the punch stopped the chain of compliments. He looked forward to getting to know her better as soon as they straightened up a bit. As he was trying to remember everything he knew about her, he decided it wasn't important. What was important was to make rice.

Today was the first day of the opening of the "committee" lakes. Luke had a car, canoe, knockers, pole, and a new partner. He only had to find lunch and sacks for the green rice. They were signed up for Dead Fish Lake and he didn't want to miss that first run through the ripe, untouched rice.

As he drove to the store to score the sacks and lunch, Dolly woke up and asked for a smoke. His crushed pack still had some left. As they lit up he noticed, she really is pretty. He shared his beer with her as she began to pull herself together. She took his rearview mirror to comb her hair, inspect her teeth, cough, and blow her nose. She could complete her toilet at the bathroom in the store.

His headache was subsiding down to a dull throb as they pulled into the store parking lot. He saw a couple of cars he recognized and his cousin's truck. He reached for his wallet and suddenly realized he had been jackrolled as he slept on the couch. There must have been someone still awake when he went to sleep.

Dolly handed him his wallet and winked. She said she took it to protect his bankroll. He was beginning to like this woman more and more. His $11.00 was still there.

For the car, Luke bought gas and a quart of oil and added air to that right rear tire that had a slow leak. For themselves, he got lunch materials and smokes. The store owner gave him a couple of sacks because he had promised to sell his rice there.

As he drove to the lake, memories of past ricing seasons came to him. His earliest memories were of playing on the shore while his parents were out ricing. He knew the people enjoyed ricing and there were good feelings all around. The years seemed to melt from people. Grandparents moved about with a light step and without their canes. Laughing and loud talking broke out frequently. The cool crisp morning air, the smell of wood smoke, roasting meat, and coffee were all part of these early childhood memories.

When he grew older, his responsibilities increased. He took care of his brothers and sisters. He cleaned the canoes and rice boats of every last kernel of rice. He learned how to make rice poles and knockers. He learned how important ricing was to the people.

His thoughts were jolted back to the present when his left rear tire blew out. He thought, "Gee that was my good tire too." As the car lurched to a stop, he knew he didn't have a spare 'cause he hocked it last week. He took the offending tire off because he knew he'd have to eventually. He looked down the road.

He saw one of his uncles coming down towards the rice landing. When his uncle saw what was wrong, he stopped and opened his trunk. He had a snow tire in there that still had some air. Luke shared a smoke and a beer with his uncle as he put the good tire on. As they sipped and gossiped, Luke found out that his uncle was not planning on ricing this year but was just going down to the landing to see if anyone needed a partner. His uncle was known as a good knocker so finding a partner was not going to be a problem.

The tire change made them late and they had to park way back on the road. The canoe and all their stuff had to be packed a long way to the lake. Everyone else was lined up ready to go.

The ricers laughed as he was getting in the water because his face showed what he'd been doing the night before. In the face of all this teasing, he could hardly wait for the jokes and jibes to be aimed at someone else so he could join in the laughter. It was good to see his relatives anyway.

Dolly was standing in front of the canoe with her pole as he arranged his sacks, water jug, and, "oops," must have left the lunch in the car. They drifted out to where the eager ones were waiting to blast off. Luke was knocking because this was his partner's first year ricing. He had looked at the rice while hunting and he knew where on the lake to go.

He repeated the string of instructions to Dolly as they began. "Don't break the rice, stay out of the open water, keep the canoe moving, watch for the color of the ripe rice, if you feel yourself falling, jump in and save the rice in the canoe."

Dolly looked comfortable in the front of the canoe and they didn't feel tippy as they started off. She poled to the rice and the sound of the other ricers swishing through the rice made a nice rhythm. Luke chuckled when he heard the unmistakable sound of a knocker hitting a canoe instead of the rice. While loosening up his arm and shoulder muscles, he saw the heavy heads were hanging at just the right angle for easy knocking. The sound of rice falling in the canoe made Luke feel good.

Everywhere he looked, there was good rice. This good patch went all the way across the lake. He saw a bald eagle circling above them. He interpreted this as a good sign.

After a couple hours, Luke's stomach reminded him of the lunch in the car. The rice had bearded up pretty good and was clean and free of debris. Not too shabby, thought Luke.

They stopped to visit with his first cousin who stopped for lunch. The fried bread, deer meat, and green tea gave Luke and Dolly enough strength to finish the day. Luke thought he'd bring his cousin some of the next deer he shot.

After getting off the lake, they sacked up on the shore. Both Luke and Dolly joined in on the laughing and exaggerating as people told stories about what happened on the lake that day. The harvest was good and everyone was happy with the way the rice was falling.

This was true except for the greedy ones who wouldn't have been satisfied with seven canoeloads of rice. They would have wanted eight.

While sacking up, they laughed because one of the greedy ones tipped over right in front of the landing. The rice would grow good there next time.

At the auction, the bids were all pretty good and they ended up selling their rice to that buyer from McGregor. The buyer knew he was getting quality rice at a low price. The buyer didn't know, in addition to the good rice, he bought half a muskrat house, wet blue jeans, and some rocks with zero nutritional value.

As the sacks were being weighed, the buyer was pulling up on the sack to decrease the weight. On the other side a skinny brown hand was pulling down just a little bit harder.

The people laughed as they watched this ancient tug of war. They were familiar with this yearly ritual.

Luke and Dolly looked forward to the next day of ricing. It should be a better one because now she knew what she was doing out on the lake. "She is pretty and pretty good at ricing," thought Luke.

Yup, ricing was here again.

IN A SMALLER WORLD

Douglas Wood

We huddled in the insufficient shelter of a tiny island. A boulder pile and a few small spruces broke the force of the wind and rain as we turned our backs against the nor'easter that was howling across three miles of open lake. It was decision time.

The storm had come up fast and was still rising. The little island offered no place for eight people to camp—barely enough room for the makeshift shelter where we now gathered. We could wait for the storm to pass, a day, maybe two or three; but we were cold and wet, and already behind schedule. Or, we could get in the canoes and head back the way we had come—going with the waves across the open channel, hoping to find a decent camp on the west shore. There we'd be in the teeth of the wind, and still further behind schedule. The third choice was to go on, bucking the storm and heading for the lee of the east shore.

Feet shuffled. Eight pairs of eyes scanned the gray

horizon. Options were discussed, fears expressed. The decision was made. We'd go.

Five minutes later the world was a thrashing maelstrom of wind and waves, of screaming muscles and muffled curses. To either side of me the other two boats leaped and pitched. The entire universe consisted of an angry lake and four canoes, and a dark shoreline that drew slowly, ever so slowly, nearer.

When at last we had reached the lee there was a celebration, compounded of exhaustion and exhilaration, and a strange, hard-to-name feeling I've come to think of as the "shrinking of the world." It's a feeling I've known many times in a decade of guiding—sometimes summoned by such an experience of stress or adventure. In the moments following that wild crossing, the world was "full"—full of the faces of seven people I cared deeply about, seven people who felt the profound nature of their ties to one another, full of the joy of a challenge overcome and the fierce glow of triumph, full of the experience of living. At such times "the world" is easily defined and its boundaries clear. I've known the feeling at gentle times, too—around a campfire in the evening, the air hanging heavy with the fragrance of cedar smoke and a loon wailing deep in the darkness. And I've sensed it on the surface of a glassy lake at dawn, enveloped in fog. These are experiences far removed from the supercharged atmosphere of physical danger or "adventure," yet the feeling was there.

I remember a night on a northern lake, a typical evening around the campfire. Dishes and camp chores done, we told some jokes and stories, reflected on the events of the day, then lapsed into silence. The silence was bent, then broken by a sound that seemed to begin under our feet and travel up our spines. Not a word was said as the wolf howl died into echoes. Glances were exchanged, and for a few moments our minds reached into the darkness and brushed against . . . mystery.

The shrinking of the world has to do with mystery, partly because it involves a paradox. At the same time the world shrinks, it expands. In any wilderness journey, the leaving behind of schedules and appointments, job and routine, in a very real way expands the world. We find ourselves opening up to the vast rhythms and tempo of the natural world—sunrises and sunsets, the songs of wind and the crashing of storm waves. All the pregnant smells, sights, and sounds to which we are so often closed in daily life are opened to us once more, each one a language and a teaching to a human organism rooted in the biological heritage of the earth. The universe itself becomes more accessible. Awareness expands.

Yet at the same time that awareness is expanding, there is the sure sense that the world is smaller, more manageable. Free from the bombardment of media images, from the chaotic rush of trivia and far-away crises and tragedies over which we have no control, we have entered a different world—a more meaningful world.

In this world, boulders and trees and animals are presences that represent forces, which are inherently symbolic and to which we are mysteriously connected.

In this world, every event is achingly real, whether tragic, troublesome, or joyous. Things either happen to me or to someone I know, someone with whom I'm interacting. A fish hook is stuck in my finger and it hurts. Or else it's in your finger and I hurt for you, and worry about what it means to the group. We have to do something about it. There are no diversions or entertainments apart from ourselves and the real world which surrounds us. The spectator mentality of mass culture is no longer an option. Each person's actions and reactions are important, and have clear consequences for the direction and well-being of our—now shrunken—world.

This heightened awareness and shift in attitude usually takes time, and it can't be forced. Sometimes there's a particular event that cracks the mask, that opens a person up to the real world and to its participants. More often it's a gradual seeping-in process, an accumulation of rainstorms, moonrises, and silence—a personal acceptance of responsibility.

For a short time, at least, I know everyone in the world, and they know me. I can look around the campfire circle and see all their faces. In this smaller world of enhanced meaning, each of us knows that everything we do or say is important. The experience is both wonderfully empowering and humbling.

The image of the circle is especially strong in evoking the dimensions of life, whether the human circle of community or the great circle of the universe. Once, on a trip far to the north, I was feeling a vague frustration. Our trip had begun well; we had been out for four days, yet I somehow felt I wasn't really there, hadn't quite "arrived." Far down the lake we could see a huge granite promontory looming high over the rest of the country. As we drew closer we saw that it was strangely beautiful, naturally terraced all the way to the top. Something pulled me, and I asked the others to go on while my partner and I climbed to the top. Once there we could see all the wild terrain over which we'd come—the stairstep rapids, the dark lakes—and what lay before us. More than that, we could turn 360 degrees and see the great circle of that wild and lonely land all around us. Suddenly I was there. The ground beneath my feet was home and frustration was gone.

The Native Americans, when taking the sacred pipe, would first offer it, stem up, to the four cardinal directions, then to Father Sky and Grandmother Earth. In such a frame of mind, one is touching the world, acknowledging at once a unique separateness from, and complete interconnectedness with, all of life. The vastness of the universe has become personal.

I remember a night, around another dying campfire, when suddenly faces were illuminated and the entire lake was lit up. A meteor streaked over our heads, trailing a tail of fire half as long as the sky. For an instant there was fear. Then awe. Then an awareness that we had been touched by a cosmos incredibly vast and mysterious. In the context of this awareness, our circle seemed infinitely larger, and—somehow—smaller. Later I went down to the rock ledge where we'd stored the canoes. They lay there like four sleek shadows, reminders of the small size and vulnerability of our group—and of groups like us throughout human history, who had risked everything to explore and travel the mysterious world around them. Now that search has led to the stars, to the realm of the meteors. I went back and rejoined the circle. They had laid another log on the fire.

As I reflect on the experiences of scores of wilderness trips, I've come to think of the concept of the "smaller world" not as an aberration or some kind of artificial state of mind; but rather as a reconnecting with the basic realities of the human condition, and of our past. It is a past steeped in the heritage of

the cave, the hunting party, the council ring. It involves not ano-
nymity, but knowing, and being known; it implies sharing,
trust, and interdependence.

Such human memories and needs run very deep,
though they often seem buried by the collective nonsense and
despair of the age in which we live. Yet sometimes, beneath the
stars and the shadows of trees, as the woodsmoke rises and firelight dances on a circle of faces, the feelings rise strong once
more, and the world is again small, and beautiful.

CONTRIBUTORS

Tom Anderson is a naturalist at the Lee and Rose Warner Nature Center near Marine on St. Croix, which is administered by the Science Museum of Minnesota, and has written extensively about the flora and fauna of northern Minnesota. His recent book, *Learning Nature by a Country Road*, is a collection of "heart-warming and honest seasonal essays, childhood reflections, folk lore, and naturalist's observations."

Henry Beston was a New Englander whose most famous work, *The Outermost House*, has been named "one of the ten best nature books in the English language." Never a North Writer in the parochial sense of this collection, he had a lifelong scholar's fascination with the legends of northern Indian tribes in the westward movement of history through the Great Lakes region.

Next to Sigurd Olson, **Les Blacklock** is perhaps the best known and most lavishly awarded of Minnesota's celebrators of north country. In the past year he was presented with an honorary Ph.D. by Northland College, in Ashland, Wisconsin; the Borrel Award, the Association of Interpretive Naturalists' top award for outstanding accomplishment; and the Keeper of the Dream Award from the YMCA at Camp Sig Olson, which is named for his wilderness mentor and close friend. *Ain't Nature Grand, Meet My Psychiatrist, Minnesota Wild, Our Minnesota* (with his wife, Fran), and *The Hidden Forest* (with Sig Olson) are perhaps his most popular books.

Michael Dennis Browne is one of Minnesota's most prolific and acclaimed poets whose work has won him fellowships from the National Endowment for the Arts and the Bush Foundation. He is a professor of English at the University of Minnesota and has written several books of poetry, including *Wife of Winter, The Sun Fetcher*, and *Smoke from the Fires*. A fourth collection, *You Won't Remember This*, is to be published by Carnegie-Mellon University Press in 1992. With composer Stephen Paulus, he has written several operas, choral pieces, song cycles, and children's songs.

Two generations ago, **"Pipestone" Cary** was known along the Boundary Waters canoe trails as one of the early canoemen and protagonists of the airplane ban and a roadless area. In St. Paul he was better known as Evan Cary, vice-president and treasurer of Gould National Battery Company. His book *Whither Away?*, the log of his many canoe camping experiences, was published in 1949.

Heart Warrior Chosa is a writer and Native American activist who was a candidate in the 1990 Minnesota gubernatorial election. She lives in a cabin near Ely, and in her contribution to this selection, she recalls the childhood delights of many summers spent at the family cabin on Basswood Lake.

Richard Coffey, a native of Rochester, writes for and publishes the *Pine County Courier* in Pine City, as well as owns and publishes *Minnesota Flyer* magazine. When he and his wife, Jeanne, moved north, they built a home in one of the more remote sections of

Pine country, where Coffey wrote the book *Bogtrotter*, which chronicles the joys, hardships, and misadventures of that project.

One of the most widely read nature writers in Minnesota, **Sam Cook** has an artistic eye for the minutiae of outdoor living. He has written on outdoor topics for the *Duluth News-Tribune* for eleven years, and his assignments have taken him to the North Pole, on dogsledding trips to Great Slave Lake, and canoe trips down God's River to Hudson's Bay. Two collections of his essays and stories have been published: *Up North* (1986) and *Quiet Magic* (1988).

William O. Douglas was a man of extraordinary vision and accomplishments. Teacher, lawyer, environmentalist, writer, and justice of the United States Supreme Court, he was born in Maine Township in Ottertail County, where his father was a home missionary of the Presbyterian church. Throughout his life Douglas was a passionate advocate of wilderness values and wrote many books about his travels in far places. *Of Men and Mountains*, *Beyond the Himalayas*, and *North from Malaysia* are perhaps the best known.

Michael Furtman is a writer, editor, and photographer who lives on the North Shore of Lake Superior and has written two popular books on northern Minnesota: *Boundary Waters Fishing Guide* and *A Season for Wilderness*, the latter the result of a summer he and his wife, Mary Jo, spent in the Boundary Waters Canoe Area Wilderness as volunteer rangers. His new book, *On the Wings of the North Wind*, chronicles his three-month journey following migrating waterfowl from north-central Saskatchewan to the bayous of Louisiana.

Carl Gawboy grew up in an Ojibway-Finnish family on a homestead near Ely. He is a well-known artist of the Lake Superior region as well as a writer, and teaches bilingual studies at the College of St. Scholastica in Duluth. He and his wife, writer Lynn Maria Laitala, and their daughter, Anna, live near Bennett, Wisconsin.

Ted Hall had a common newspaper journalist's dream: he wanted to edit and publish a small-town newspaper of his own. After stints with several East Coast metropolitan dailies, *Time* magazine, and NBC News, he settled in Ranier (pop. 237), just east of International Falls, where he publishes the *Rainy Lake Chronicle* and operates a printing company. His delightful and sometimes irreverent columns have been collected in a book, *Rainy Lake Chronicle: Warm Tales from a Cold Northland Town That Time Has Not Forgotten.*

Living in a ghost town is just one of the interesting things about **Joanne Hart**. A poet of extraordinary sensitivity and perspectives, essayist, and self-described "retired mother of eight children," she lives on the Grand Portage Indian Reservation, nine miles north of town in the deserted village that used to mark the border crossing into Canada. Three books of her poems, *I Walk the River at Dawn*, *In These Hills*, and *The Village Schoolmaster*, are reflections of life among her Ojibway neighbors. A new collection, *Witch Tree* (with artist Hazel Belvo), was published in 1991.

At the top of any lover of the north country's "best job" list would be the ones held by **Ellen Hawkins** and her husband, Dick Brandenburg. They are wilderness canoe rangers, working out of the U.S. Forest Service's Tofte District Ranger Station. Ellen is also a part-time teacher at the Birch Grove Elementary School in Tofte, and is producing a video and written material for wilderness users. Several years ago, she and Dick took a sick male timber wolf into their cabin and tried to nurse it back to health. "Wolf at the Window" is the story of that unlikely adventure as it appeared in *Audubon* magazine.

Like his Gunflint Lake neighbors, Justine Kerfoot and the late Helen Hoover, **John Henricksson** writes about the cultural and natural history of northeastern Minnesota and his work appears in *Minnesota Monthly, Lake Superior, Sports Afield, Encounters, Outdoor America*, and the Sunday *St. Paul Pioneer Press*. His recent book, *Rachel Carson*, traces the beginnings of the environmental movement in America.

In 1947, **Helen Hoover** and her husband, Ade, who illustrated her books, moved from the clamor of postwar Chicago to a remote cabin on Gunflint Lake. Here she wrote many magazine articles, books, and children's stories about her adopted wilderness country, among them *Gift of the Deer*, which was chosen as a *Reader's Digest* condensed book in 1966. Her other nature classics from the Gunflint region are *The Long Shadowed Forest*, *A Place in the Woods*, and *The Years of the Forest*.

Florence Page Jaques, wife of the renowned nature artist, Francis Lee Jaques, wrote two north country classics: *Canoe Country* and its winter companion piece, *Snowshoe Country*. The latter received the John Burroughs Medal in 1942. Her books *Birds across the Sky*, *As Far as the Yukon*, and several travel books and magazine articles, were the result of her travels with her husband.

No one knows the Gunflint Trail region of northern Minnesota as well as **Justine Kerfoot,** who has walked its trails, canoed its waters, and watched over it with love and concern since 1927. She owned and operated Gunflint Lodge and Gunflint Outfitters from 1930 to 1978 and then became a county commissioner, newspaper columnist, Republican party national delegate, world traveler, and author. Her two books, *Woman of the Boundary Waters* and *Gunflint: Reflections on the Trail*, tell the story of her sixty-four years on Gunflint Lake.

Lynn Maria Laitala, a native of the Ely area, is a writer, historian, and college teacher who raises horses on a small farm near Bennett, Wisconsin, where she lives with her husband, Carl Gawboy, and daughter, Anna. Her writing projects are primarily history, current events, and fiction.

After several grinding years of study for his Ph.D. in American literature at the University of Utah, **Paul Lehmberg,** a native Minnesotan and former staff member of the Outward Bound School near Ely, returned to the wilderness to build a cabin just across the Ontario border in the Quetico Provincial Park. Currently, Lehmberg is a professor of English and director of the graduate writing program at Northern Michigan University at Marquette, where he continues to produce essays and reviews.

Peter Leschak, who lives at Side Lake near Chisholm, has written essays, reviews, and satires that have been published in a wide variety of publications including the *New York Times, Harper's* magazine, *Minnesota Monthly, New North Times,* and *Boundary Waters Journal.* His book *Letters From Side Lake* was published by Harper & Row in 1987, and his latest book, *Bear Guardian,* is a collection of north woods tales and campfire stories.

Author, educator, and naturalist, **Mike Link** is the director of the Audubon Center of the Northwoods at Sandstone, and has taught ecology at Northland College in Ashland, Wisconsin, and the University of Minnesota at Duluth. He has written eight books and dozens of magazine articles on the natural world and has edited a three-volume collection of the works of Sigurd Olson. His most recent book, *Love of Loons,* was coauthored with his wife, Kate Crowley, also a naturalist at the Audubon Center.

Essayist **Matthew Miltich** was a Marine Corps sergeant during the Vietnam War and then became a "born-again primitive" in northern Minnesota, where he teaches English at Itasca Community College and farms in Wabana Township. He, his wife, Loree, who is editor of *New North Times,* and their children raise much of their own food and are as self-sufficient as possible. Miltich's essays have been featured in *Loon Feather, Whitetales, Waterfowler's World,* and on public radio.

Although currently a resident of St. Paul, **Judith Niemi** claims a northern Minnesota identity. Born and raised in Eveleth on the Iron Range, she has a cabin on Lake Vermilion and has guided canoe and winter camping trips since 1977. Her business, Women in the Wilderness, organizes all-season adventure tours for women to Canada, Labrador, Alaska, and Nepal. She is an adjunct faculty member at Metropolitan State University, Mankato State University, and St. Mary's College. The author of numerous books, travel articles, book reviews, and essays, Niemi won the 1990 Page One Award from the Society of Professional Journalists.

Jim Northrup, Jr., Anishinabe writer, poet, and storyteller, and his wife, Patricia, live on Fond du Lac Reservation new Duluth. A Marine Corps veteran of the Vietnam conflict, he is a writer in residence for the COMPAS Program and mentor for the Loft Inroads Program. His work is published in several anthologies, including *Touchwood, Stiller's Pond*, and *Shorelines*. He recently won the Lake Superior Contemporary Writer's series and took first place in the Native American Press Association's feature-writing contest.

One of the Minnesota's most illustrious historians, **Dr. Grace Lee Nute** was state archivist and mentor to many of the scholars now on the staff of the Minnesota Historical Society. Her best-known book, *The Voyageurs*, became the foundation of much of the Society's publications program. Her other major contributions to the history of Minnesota's north country are *Rainy River Country, The Voyageur's Highway*, and *Caesars of the Wilderness*.

For several years a naturalist at the Wolf Ridge Nature and Environmental Learning Center near Isabella, **Denny Olson** has parlayed his scientific knowledge, acting ability, and good humor into a multifaceted career that includes school presentations, national conferences, and special events in forty-seven states appearing before 1.2 million people. His *Season of Stillness* is part of a series of seasonal essays published in the *Boundary Waters Journal*.

To the lovers of the canoe country, **Sigurd Olson** needs no introduction. Canoe guide, college teacher, fighter for wilderness, author, and philosopher, Olson's name will be forever synonymous with the Quetico-Superior. He is the man they all called *Bourgoise*, the Leader. The consummate "environmental statesman," Olson was responsible for leading the two-decade-long battle to preserve the Boundary Waters Canoe Area Wilderness. Mike Link has edited much of his popular writing and many of his lesser-known essays and fiction into the three-volume *Collected Works of Sigurd F. Olson*. His best-known books are *Singing Wilderness, Runes of the North, Reflections of North Country, Open*

Horizons, Listening Point, The Lonely Land, Song of the North, The Hidden Forest, and *Wilderness Days.*

They call **Lynn Rogers** "the Bearman." He is one of the foremost authorities on the black bear, and his credentials include a doctorate from the University of Minnesota in wildlife ecology and the directorship of the black bear research team at the U.S. Forest Service Experiment Station at Ely. He is also the author of more than eighty scientific papers on *Ursus americanus.* His wildlife pictures have appeared on the covers of *National Geographic, Field and Stream, Outdoor Life, National Wildlife, Smithsonian, Sports Afield,* and *Natural History* magazines.

In many ways **Calvin Rutstrum** was the quintessential Northman. For over fifty years he roamed from the St. Croix River to the Arctic—trapping, building cabins, outfitting, guiding, and writing in a salty and sometimes controversial style. A self-educated person, Rutstrum made important contributions to both literature and the sciences. There were no roads or signposts in most of the country he traveled, so he invented his own navigational and surveying equipment, some of which is still in use today. His books are for practical outdoorspersons: *North America Canoe Country, Chips from a Wilderness Log, Paradise below Zero, How to Keep from Getting Lost on Land or Sea, The Wilderness Cabin, Once upon a Wilderness,* and *The Wilderness Route Finder.*

Mark Sakry brings some incongruous skills and a varied background to his present, contented life in the Cloquet State Forest near Brimson. A self-described "escaped technical writer and copy editor," in his current "reincarnation" Sakry's primary interests are outdoor education, wild foods, winter camping, environmental activism, teaching, and writing. *Sports Afield, Lake Superior, Backpacker, Boundary Waters Journal,* and the Duluth *News-Tribune* carry his work, which stresses reader participation.

Milt Stenlund is a native of Ely, a resident of Grand Rapids, and the retired regional game manager for the Minnesota Department of Natural Resources. He has written hundreds of technical articles including the benchmark "Field Study of the Timberwolf" in 1953, the first comprehensive field research ever done on the Minnesota timberwolf. Absorbed in the history of northern Minnesota, Stenlund has recently published two books on the Ely area: *Burntside Lake: The Early Days* and *Ghost Mines*. His book *Popple Leaves and Boot Oil* is the memoir of his thirty-five-year career as game manager, wildlife biologist, and administrator of conservation programs in northern Minnesota.

Robert Treuer has come to love the freedom and beauty of northern Minnesota as only one who has fled the Nazi holocaust can understand. He lives on the cutting edge of ethical dilemmas, has become part of an Indian family and community, has been a labor union activist, an early civil rights fighter, and a tree farmer, and has written passionately about the issue of human being/land relationships. Two of his popular books are *The Tree Farm* and *A North Country Window*. A third, *Voyageur Country*, tests the thesis that ravaged land can be returned to a viable ecosystem.

From his experiences living in a cabin near Ely, **Jim dale Vickery** wrote the essay "The Land Is Alive with Wolves," for *Audubon* magazine. This piece was later cited in Atwan-Dillard's *Best American Essays*, and was also reprinted in Audubon's 1989 *Nature Yearbook*. His book *Wilderness Visionaries* profiles Robert Service, John Muir, Bob Marshall, Calvin Rutstrum, Sigurd Olson, and Henry David Thoreau in the context of North America's wilderness preservation movement. A new collection of essays, *Open Spaces*, chronicles his wilderness wanderings through Alaska, Canada, and northern Minnesota. Vickery guided canoe trips in the Quetico-Superior and also is a seasonal National Park ranger in the Apostle Islands.

Troubadour **Douglas Wood** is best known as a musician, storyteller, writer, naturalist, and wilderness guide. He combines art, science, and environmental education through the medium of music in his public performances and on cassette tapes of *Earthsongs* for the Science Museum of Minnesota. He is much in demand as a conference speaker on the theme of wilderness and the human spirit, and frequently appears on Twin Cities radio and television.

PERMISSIONS

Tom Anderson, "The Passing of Thrushes," from the *Chisago County News*, September 10, 1987, copyright the author, by permission; Henry Beston, "The Golden Age of the Canoe," from *The Great Lakes Reader*, ed. Walter Havighurst, first published 1941, copyright Collier Books/Macmillan, 1978, by permission; Les Blacklock, excerpt from *Meet My Psychiatrist*, Voyageur Press, 1978, copyright the author, by permission; Michael Dennis Browne, excerpt from "North Shore," copyright the author, by permission; E. F. Cary, "Whither Away," from *The Great Lakes Reader*, ed. Walter Havighurst, first published 1941, copyright Collier Books/Macmillan, 1978, by permission; Heart Warrior Chosa, "Basswood Lake," from *Heart of Turtle Island: A Trilogy*, book 1, "Seven Chalkhills," copyright the author, by permission; Richard Coffey, excerpt from *Bogtrotter*, published by Dorn Publications, copyright 1982 Dorn Books and the author, by permission; Sam Cook, "Caterpillar Nap," from *Up North*, Pfeifer-Hamilton, 1986, copyright Pfeifer-Hamilton and the author, by permission; William O. Douglas, "Quetico-Superior," from *My Wilderness–East to Katahdin*, Doubleday & Company, 1961, permission applied for; Michael Furtman, "Never Throw Stones at a Maymaygwashi," from *A Season for Wilderness*, NorthWord Press, 1989, copyright the author, by permission; Carl Gawboy, "Christmas at Birch Lake," from *Sampo: The Magic Mill*, New Rivers Press, 1989, copyright the author, by permission; Joanne Hart, "Why I Live in These Hills," from *In These Hills*, Women's Times Publishing, 1981, second printing 1986, copyright the author, by permission; Ellen Hawkins, "The Wolf at the Window," from *Audubon* magazine, November 1988, copyright the author, by permission; John Henricksson, "New Sport for an Old Hunter," from *Outdoor America* (The Izaak Walton League of America Magazine), Winter 1989, copyright the author, by permission; Helen Hoover, "King Weather," from *The Long-Shadowed Forest* by Helen Hoover, illustrated by Adrian Hoover, by permission of W. W. Norton & Company, Inc., copyright 1963 by Helen Hoover and Adrian Hoover; Florence Page Jaques, excerpt from *Canoe Country*, copyright 1938, the University of Minnesota Press, 9th printing 1979; Justine Kerfoot, "A Dog Team Trip," from *Woman of the Wilderness*, Women's Times Publishing, 1986, copyright the author, by permission; Lynn Maria Laitala, "Winter Trip," from *Sampo: The Magic Mill*, New Rivers Press, 1989, copyright the author, by permission; Paul Lehmberg, "The Order of Things," from *In the Strong Woods: A Season Alone in the North Country*, St. Martin's Press, 1980, copyright the author, by permission; Peter M. Leschak, "First Ice: Dancing on Water with a World at Your Feet," from *Minnesota Monthly*, February 1989, copyright the author, by permission; Mike Link, "Rivers" and "Rocks and Lichens," from *Boundary Waters Canoe Area Wilderness*, published by Voyageur Press, 1987, copyright the author, by permission;